The ESRI® Guide to GIS Analysis

Volume 1: Geographic Patterns & Relationships

Andy Mitchell

ESRI Press

PUBLISHED BY
Environmental Systems Research Institute, Inc.
380 New York Street
Redlands, California 92373-8100

Environmental Systems Research Institute, Inc.
The ESRI Guide to GIS Analysis
Volume 1: Geographic Patterns and Relationships
ISBN 1-879102-06-4

Contents

Preface

Spatial analysis is where the GIS rubber hits the road—where all the hard work of digitizing, building a database, checking for errors, and dealing with the details of projections and coordinate systems finally pays off in results and better decisions. But spatial analysis has often seemed inaccessible to many users—too mathematical to understand, too difficult to implement, and lacking in good textbooks and guides. Here at last is the ideal book, written by Andy Mitchell and based on ESRI's vast experience with applications of spatial analysis to a host of real problems. The book covers every area of GIS application, so readers will find examples that relate directly to their own concerns, whether they be in hydrology, transportation, or regional planning. The organization is intuitive, with sections on all of the major forms of simple spatial analysis. As noted in chapter 1, 'Introducing GIS analysis,' ESRI plans to follow and build on this with a second more advanced book, which will cover some of the more complex methods.

We tend to think of spatial analysis as something different from mapping—and substantially more sophisticated. The phrase "just a mapping project" is often heard in GIS circles, and carries with it the implication that if sophisticated GIS software is used only to display data in visual form that somehow it is being underutilized. In fact the earliest GIS—the Canada Geographic Information System—had no display capabilities at all in its original design, and could only produce numerical output in table form. One of this book's most valuable contributions is to show how mapping and analysis are intimately linked, and how we gain most from GIS when we combine carefully designed visual display with numerical summaries. The book includes abundant examples in color of the use of GIS for visual display. As the author points out, analysis does not have to involve complex mathematical operations, but begins in the human mind as soon as the map is in view, because the eye and brain are enormously efficient at detecting patterns and finding anomalies in maps and other visual displays. GIS works best when the computer and the brain combine forces, and when the GIS is used to augment human intuition by manipulating and displaying data in ways that reveal things that would otherwise be invisible.

This book will appeal to GIS users in all areas of GIS application. It will be invaluable reading for people encountering GIS for the first time, and wanting to see where its real power lies. It will make an excellent textbook for courses in GIS in high schools, community colleges, and undergraduate programs, and as a supplement for practical work. This is not a software manual, and it rightly avoids tying to any particular brand or version of GIS. So although its origins are in ESRI, it should be valuable to users working with other brands of GIS, and will be useful to professionals who are concerned about interoperability across different brands, because of its focus on the basic principles of simple spatial analysis rather than on any one software product. ESRI has done the field a great service by sponsoring and publishing this book.

Michael F. Goodchild
National Center for Geographic Information
and Analysis
University of California, Santa Barbara

Acknowledgments

Many people contributed their knowledge to this book. The GIS professionals who have presented maps and papers at the annual ESRI User Conference over the years provided the real-world context for many of the concepts and methods presented in the book. Some of their maps appear in the 'Map gallery' section of each chapter; the individuals and organizations who created them are listed in the section, 'Map and data credits.' Others, mainly from academia, contributed through their publications, which are listed in the section, 'Where to get more information.'

Several people at ESRI offered their expertise, both through their review of the manuscript, and in informal discussions; these include Clint Brown, Steve Kopp, Charlie Frye, Mike Minami, Jay Sandhu, and Damian Spangrud. Several others outside of ESRI also offered valuable comments on the book: Dr. Michael Goodchild of the University of California at Santa Barbara, who also wrote the preface; Dean Angelides of Vestra Resources; and Dr. Duane Marble, who reviewed an early draft of several of the chapters.

A number of organizations provided the data used to create the map examples in the book. They are listed in the 'Map and data credits' section. Several people at ESRI—including Hugh Keegan, Mark Smith, Lee Johnston, and John Calkins—made datasets available for the examples.

Many at ESRI Press and other departments at ESRI helped with the production of the book. Michael Karman edited the book, as did Jennifer Wrightsell, who was also largely responsible for the final design. Michael Zeiler and Youngiee Auh also contributed to the design of the book. Amaree Israngkura designed the cover. Christian Harder managed the publication of the book, and Cliff Crabbe oversaw the print production.

Finally, special thanks to Jack Dangermond and Clint Brown, who recognized the value of publishing a guide to GIS analysis, and provided the support for writing it.

Introducing GIS analysis

GIS analysis lets you see patterns and relationships in your geographic data. The results of your analysis will give you insight into a place, help you focus your actions, or help you choose the best option.

In this chapter:

- *What is GIS analysis?*
- *Understanding geographic features*
- *Understanding geographic attributes*

Geographic information system (GIS) technology is now about 30 years old. However, for the most part, people are still using it only to make maps. GIS can do much more. Using GIS for analysis, you can find out why things are where they are and how things are related. By learning to use GIS for analysis, you can get more accurate and up-to-date information, and even create new information that was unavailable before. Having this information can help you gain a deeper understanding of a place, make the best choices, or prepare for future events and conditions.

So why aren't more people using GIS for analysis? One reason is that GIS use has only recently become widespread, so it's still new to many people. Because of this, many organizations are only now finishing building their GIS databases (a process that took a long time in the past but is becoming faster with the huge amount of geographic data now available). Another reason is that using GIS for analysis has been difficult and cumbersome. Now, new easy-to-use software employing graphic interfaces is removing that obstacle. A third reason more people aren't using GIS for analysis is that they don't know what they can do with GIS beyond making maps and creating reports. Or, if they do, they don't know how to go about it. While geographic data is becoming more widely available and GIS software is becoming easier to use, to do effective GIS analysis you still need to know how to structure your analysis and which tools to use for a particular task.

That's where this book comes in. You may not be aware of it, but if you make maps today, you are in effect doing analysis already. One of the goals of this book is to help you build better maps—maps that clearly and accurately present the information you need from your data. We also want to introduce you to some of the basic analysis concepts and tasks that—while useful in their own right—are the building blocks for more advanced analysis.

In this book, we've identified the most common geographic analysis tasks people do every day in their jobs:

- Mapping where things are
- Mapping the most and least
- Mapping density
- Finding what's inside
- Finding what's nearby
- Mapping change

The book is organized in three parts. In this chapter, you'll learn what GIS analysis is, and what it can do for you. You'll also review some basic GIS concepts: what geographic data is and how it's stored, and more about data values, their use, and interpretation. Chapters 2 through 4 present key map-building concepts. They focus on ways of presenting geographic data to best see the patterns of how things are distributed. The later chapters focus on map query and map-based analysis tasks that let you look at geographic relationships.

In the next decade, the use of GIS analysis will grow. A new type of user will emerge—the spatial scientist. A significant number of GIS users will emerge as advanced modelers. Our goal is to help you expand your analytical GIS skills and sophistication. To do that, ESRI plans to add another book to this series covering more advanced analysis concepts and methods.

GIS analysis is a process for looking at geographic patterns in your data and at relationships between features. The actual methods you use can be very simple—sometimes, just by making a map you're doing analysis—or more complex, involving models that mimic the real world by combining many data layers.

The chapters in this book follow the process you go through when performing an analysis.

Frame the question

You start an analysis by figuring out what information you need. This is often in the form of a question. Where were most of the burglaries last month? How much forest in each watershed? Which parcels are within 500 feet of this liquor store? Being as specific as possible about the question you're trying to answer will help you decide how to approach the analysis, which method to use, and how to present the results.

Other factors that influence the analysis are how it will be used and who will use it. You might simply be exploring the data on your own to get a better understanding of how a place developed or how things behave; or you may need to present results to policy makers or the public for discussion, for scientific review, or in a courtroom setting. In the latter cases, your methods need to be more rigorous, and the results more focused.

Understand your data

The type of data and features you're working with help determine the specific method you use. Conversely, if you need to use a specific method to get the level of information you require, you might need to obtain additional data. You have to know what you've got (the type of features and attributes, discussed later in this chapter), and what you need to get or create. Creating new data may simply mean calculating new values in the data table (see 'Working with tables' later in the chapter) or obtaining new layers.

Choose a method

There are almost always two or three ways of getting the information you need. Often, one method is quicker and gives you more approximate information. Others may require more detailed data and more processing time and effort, but provide more precise results. You decide which method to use based on your original question and how the results of the analysis will be used. For example, if you're doing a quick study of assaults in a city to look for patterns, you might just map the individual crimes and look at the maps. If the information will be used as evidence in a trial, though, you might want a more precise measure of the locations and numbers of assaults for a given time period.

Process the data

Once you've selected a method, you perform the necessary steps in the GIS. In this book, we give you some of the concepts behind what the GIS is doing, so you can better interpret the results. We also give you some context for choosing parameters that might be required during the analysis.

Look at the results

The results of the analysis can be displayed as a map, values in a table, or a chart—in effect, new information. You need to decide what information to include on your map, and how to group the values to best present the information. You must also decide whether charts would help others easily see the information you're presenting.

Looking at the results can also help you decide whether the information is valid or useful, or whether you should rerun the analysis using different parameters or even a different method. GIS makes it relatively easy to make these changes and create new output. You can compare the results from different analyses and see which method presents the information most accurately.

As you can see, the type of geographic features you're working with affect all steps of the analysis process. Spending some time up front looking at your data—and figuring out how it can be analyzed—will make the process run smoothly. Following is a discussion and definitions of the different types of geographic data, how they're represented in the GIS, and how you'll be working with them.

You need to be aware of the different types of geographic features, different ways they're represented, and a little bit about map projections and coordinate systems.

TYPES OF FEATURES

Geographic features are either discrete, continuous phenomena, or summarized by area.

Discrete features

For discrete locations and lines, the actual location can be pinpointed. At any given spot, the feature is either present or not.

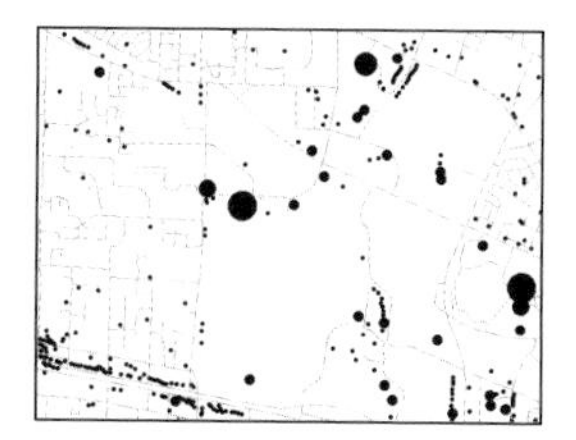

Businesses, symbolized by number of employees, are an example of individual locations.

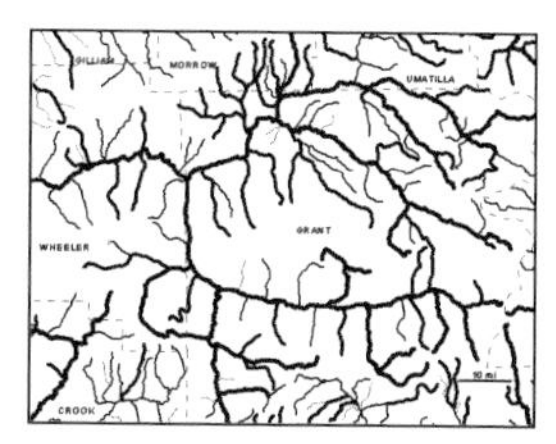

Streams are linear features.

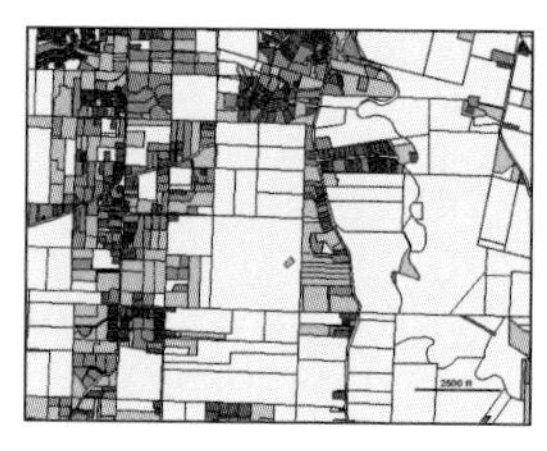

Parcels, color-coded by land value, are an example of discrete areas.

Continuous phenomena

Continuous phenomena such as precipitation or temperature can be found or measured anywhere. These phenomena blanket the entire area you're mapping—there are no gaps. You can determine a value (annual precipitation in inches or average monthly temperature in degrees) at any given location.

Continuous data often starts out as a series of sample points, either regularly spaced (such as sampled elevation data) or irregularly spaced (such as weather stations). The GIS uses these points to assign values to the area between the points, in a process called interpolation. Some noncontinuous data is treated as continuous—for a given place—in order to create maps showing how a quantity varies across the place. For example, you could create a map of land values by interpolating the center points (centroids) of all the parcels in the city.

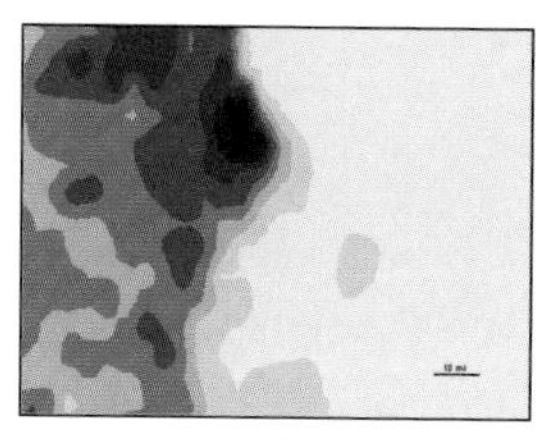

Average annual precipitation as a continuous surface

Continuous data can also be represented by areas enclosed by boundaries, if everything inside the boundary is the same type, such as a type of soil or vegetation. Of course, since the data varies continuously across the landscape, the boundaries really indicate where things are more similar than not—they're not really definitive as they are with discrete areas such as parcels, where the boundary is legally defined.

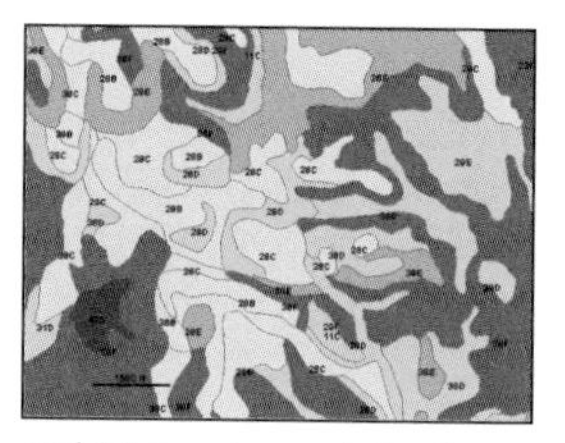

Soil types, represented using boundaries

Features summarized by area

Summarized data represents the counts or density of individual features within area boundaries. Examples of features summarized by area include the number of businesses in each ZIP Code, the total length of streams in each watershed, or the number of households in each county (obtained by summing the number of households in each census tract). The data value applies to the entire area, but not to any specific location within it.

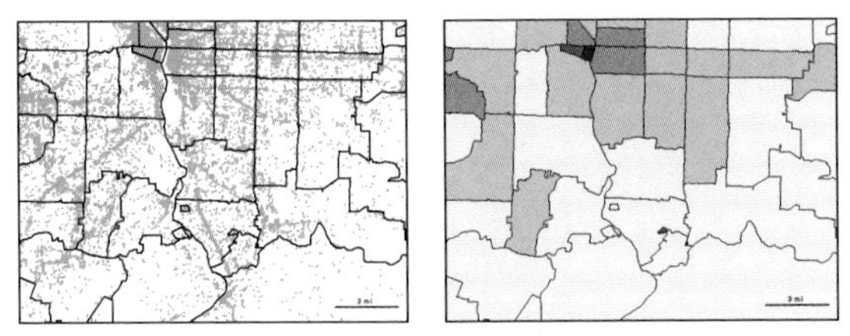

The map on the left shows business locations. The map on the right shows ZIP Codes color-coded by number of businesses.

A lot of data already comes summarized by area, especially demographic data, which consists of totals (total population, total households, and so on) or a percentage based on a category (percent over 65 years old, or percent Hispanic). Some business data is also aggregated to these boundaries or to ZIP Codes, area codes, or other boundaries.

You may have other data you want to summarize by area. If the features are already tagged with a code assigning them to the area, it's just a matter of doing statistics on the data table—for example, sum the total revenue for all businesses in each ZIP Code. That figure is then assigned to each area boundary when you join the two tables. You can then use that attribute to map the areas and look for patterns.

Harris Welco	97210
B & I Furnishings Inc	97210
Paragon Fire Sprinklers	97006
Simply Dramatic	97212
G-S Associates Inc	97210

97006	717
97007	528
97008	191
97009	245
97010	1

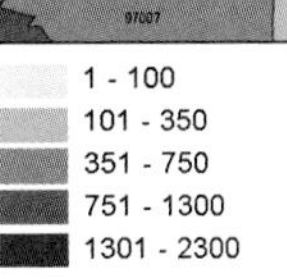

Businesses summed by ZIP Code

If the features aren't tagged with the codes for the areas by which you want to summarize them, the GIS lets you overlay the areas with the features to find out which ones are within each area and tag them with the appropriate code. You'll read more about this in chapter 5, 'Finding what's inside.'

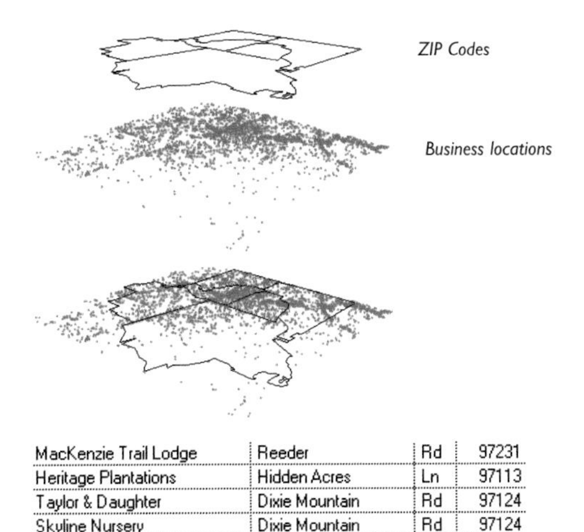

MacKenzie Trail Lodge	Reeder	Rd	97231
Heritage Plantations	Hidden Acres	Ln	97113
Taylor & Daughter	Dixie Mountain	Rd	97124
Skyline Nursery	Dixie Mountain	Rd	97124
Skyline Hills Ranch	Skyline	Blvd	97124

By overlaying ZIP Code boundaries and businesses, you can tag each business with its ZIP Code.

TWO WAYS OF REPRESENTING GEOGRAPHIC FEATURES

Geographic features can be represented in the GIS using two models of the world: vector and raster.

With the vector model, each feature is a row in a table, and feature shapes are defined by x,y locations in space (the GIS connects the dots to draw lines and outlines). Features can be discrete locations or events, lines, or areas. Locations, such as the address of a customer, or the spot a crime was committed, are represented as points having a pair of geographic coordinates.

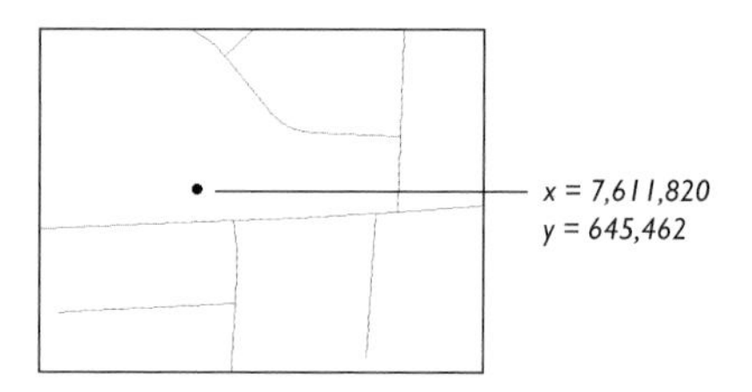

Lines, such as streams, roads, or pipelines, are represented as a series of coordinate pairs.

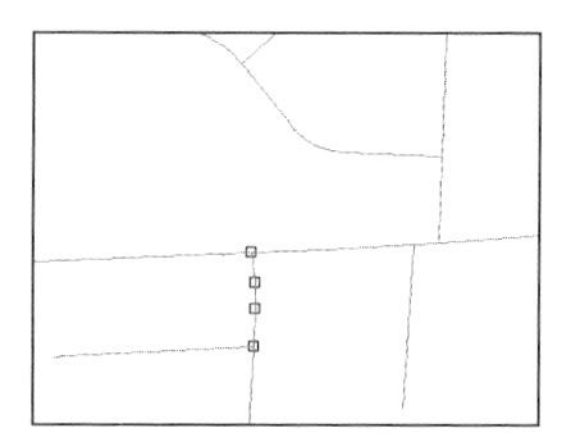

Areas are defined by borders, and are represented as closed polygons. They can be legally defined, such as a parcel of land; administrative, such as counties; or naturally occurring boundaries, such as watersheds. When you analyze vector data, much of your analysis involves working with (summarizing) the attributes in the layer's data table.

With the raster model, features are represented as a matrix of cells in continuous space. Each layer represents one attribute (although others can be attached), and most analysis occurs by combining the layers to create new layers with new cell values.

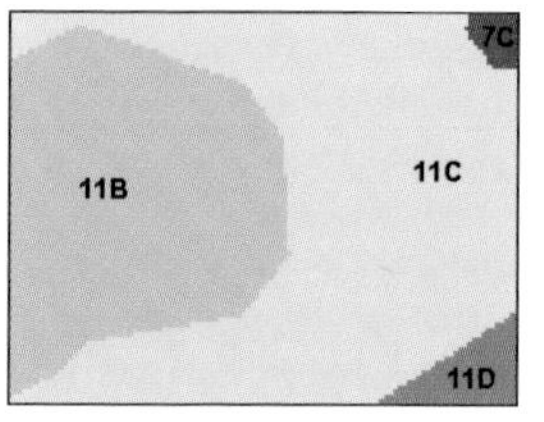

Soil types represented as a raster layer

The cell size you use for a raster layer will affect the results of the analysis and how the map looks. The cell size should be based on the original map scale and the minimum mapping unit. Using too large a cell size will cause some information to be lost. Using a cell size that is too small requires a lot of storage space, and takes longer to process, without adding additional precision to the map.

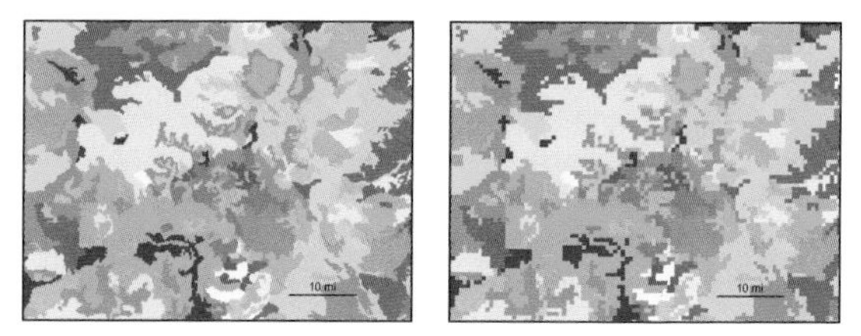

Raster layer of vegetation. A larger cell size (right map) shows the patterns, but some detail is lost.

While any feature type can be represented using either model, discrete features and data summarized by area are usually represented using the vector model. Continuous categories are represented as either vector or raster, and continuous numeric values are represented using the raster model.

Vector *Raster*

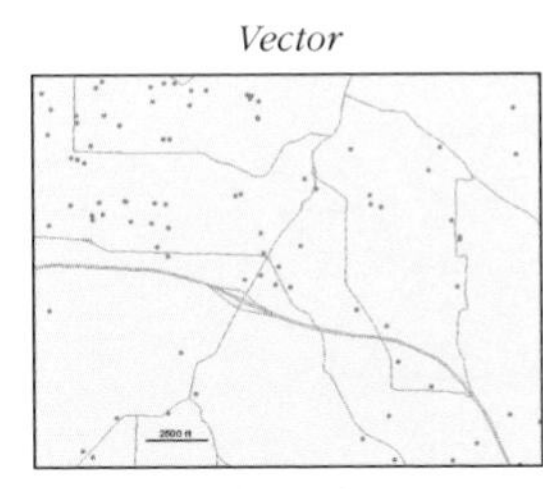

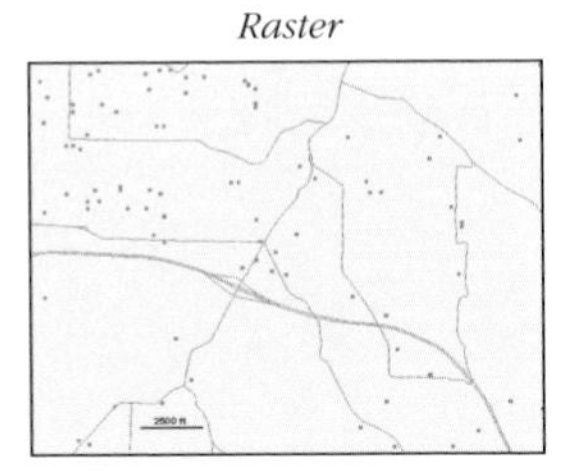

Businesses (points)

Highways (lines)

Land use (areas)

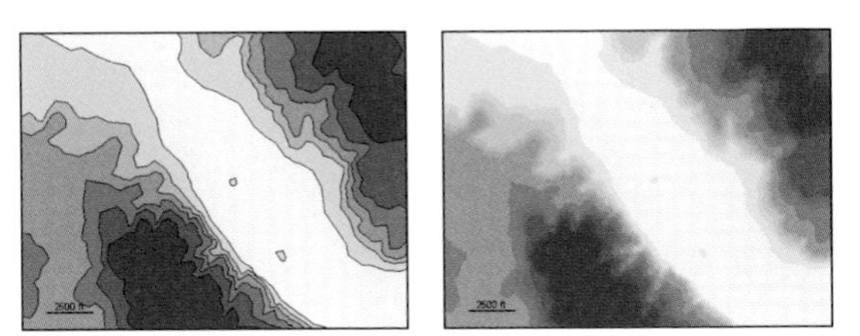

Elevation (continuous phenomena)

You might also represent discrete features as raster when you're combining them with other layers in a model, since raster is particularly good for this kind of analysis.

MAP PROJECTIONS AND COORDINATE SYSTEMS

All the data layers you're using should be in the same map projection and coordinate system. Otherwise, you won't be able to draw them on top of each other, or combine them to see relationships, such as what features are inside an area, or what features are near another feature.

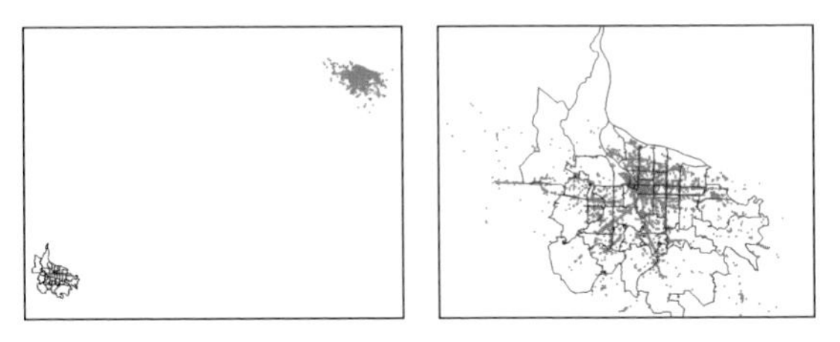

Businesses and ZIP Code boundaries in different coordinate systems (left map) and in the same coordinate system (right map)

A map projection translates the locations on the globe (which is almost a sphere) onto the flat surface of your map. All map projections distort the shapes of the features being displayed, as well as measurements of area, distance, and direction. In general, if you're mapping a relatively small area, such as a town or county, this distortion is negligible. It may be more of a concern if you're mapping a large area such as a state, country, or the entire world, because the curvature of the earth begins to come into play.

A coordinate system specifies the units used to locate features in two-dimensional space, and the origin point of those units.

If you're using an established GIS database, chances are the data you're using is already in the same coordinate system and projection. If you're collecting data from various sources, though, you'll want to check this. Several issues are involved in choosing a map projection and coordinate system, including where on the globe the area you're mapping is located, how large the area is, and whether you need precise measurements of distance or areal extent. Several of the references at the end of the book contain information on how to choose a coordinate system and map projection, and how to project your data.

Each geographic feature has one or more attributes that identify what the feature is, describe it, or represent some magnitude associated with the feature. The type of analysis you do depends partly on the type of attributes you're working with.

TYPES OF ATTRIBUTE VALUES

Attribute values include:

- Categories
- Ranks
- Counts
- Amounts
- Ratios

Categories

Categories are groups of similar things. They help you organize and make sense of your data. All features with the same value for a category are alike in some way, and different from features with other values for that category. For example, you can categorize roads by whether they're freeways, highways, or local roads, and crimes by whether they're burglaries, thefts, assaults, and so on.

Category values can be represented using numeric codes or text. Text values are often abbreviations, to save space in the table.

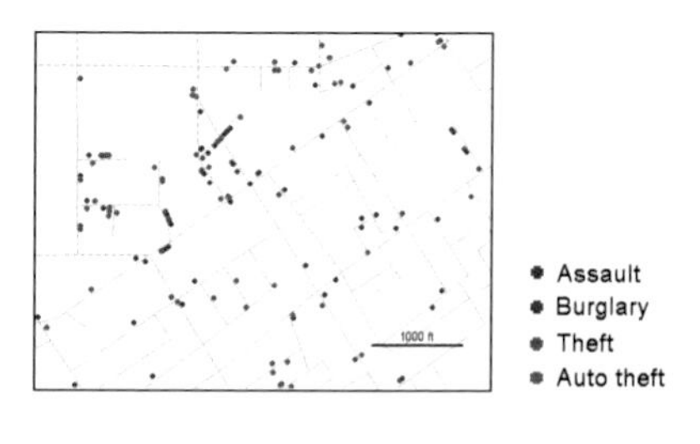

ID	Date	Type	Description
108161454	08/11/97	629	THEFT, $200-$400
107941626	07/20/97	521	BURG, UNL ENT, RES NITE
109040815	11/07/97	521.1	BURG, UNL ENT, GAR NITE
106270910	02/03/97	513	BURG, FORCED, RES UNK
109040843	11/07/97	619	THEFT, OVER $400

Crime categories are represented here using a numeric code.

Ranks

Ranks put features in order, from high to low. Ranks are used when direct measures are difficult, or if the quantity represents a combination of factors. For example, it's hard to quantify the scenic value of a stream. You may be able to state, though, that the section that passes through a mountain gorge has a higher scenic value than the section near a dairy farm.

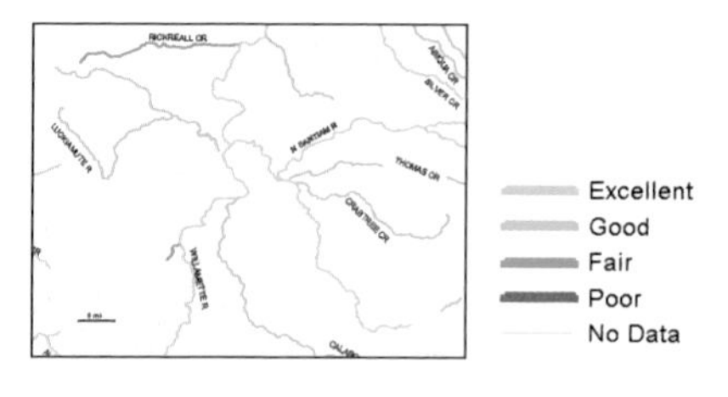

Length (ft.)	Name	Rank
79678.594	LOBSTER CR	1
22115.541	WILLAMETTE R	1
2231.341	RICKREALL CR	3
34173.461	LITTLE ABIQUA R	
165179.391	BUTTE CR	2
68918.680	LITTLE PUDDING R	

Streams ranked by recreation value

Since the ranks are relative, you only know where a feature falls in the order—you don't know how much higher or lower a value is than another value. For example, you may know a feature with a rank of 3 is higher than one ranked 2 and lower than a 4, but you don't know how much higher or lower.

You can assign ranks based on another feature attribute, usually a type or category. For example, you'd assign all soils of a certain type the same suitability for growing a particular crop.

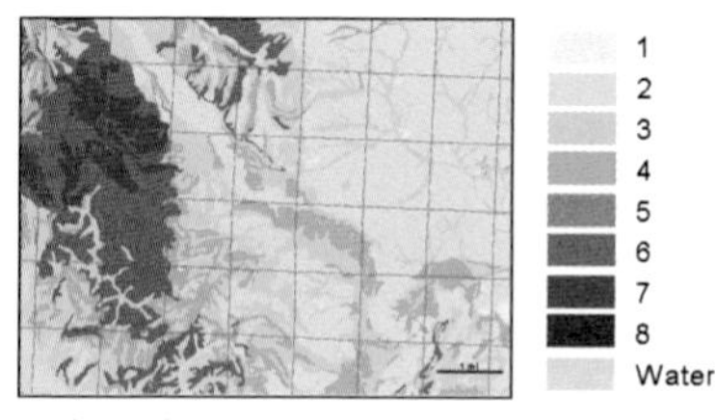

Soils ranked by suitability for growing crops

Counts and amounts

Counts and amounts show you total numbers. A count is the actual number of features on the map. An amount can be any measurable quantity associated with a feature, such as the number of employees at a business. Using a count or amount lets you see the actual value of each feature as well as its magnitude compared to other features.

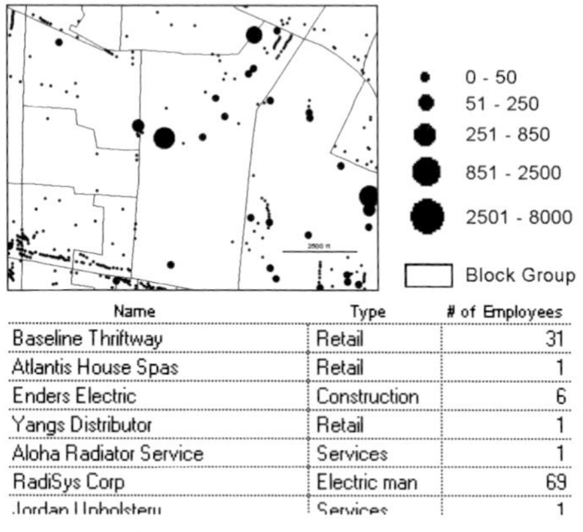

Name	Type	# of Employees
Baseline Thriftway	Retail	31
Atlantis House Spas	Retail	1
Enders Electric	Construction	6
Yangs Distributor	Retail	1
Aloha Radiator Service	Services	1
RadiSys Corp	Electric man	69
Jordan Upholstery	Services	1

Businesses by number of employees

Ratios

Ratios show you the relationship between two quantities, and are created by dividing one quantity by another, for each feature. For example, dividing the number of people in each tract by the number of households gives you the average number of people per household. Using ratios evens out differences between large and small areas, or areas having many features and those having few, so the map more accurately shows the distribution of features.

Tract	Population	Households	People per HH
003603	1606	643	2.5
0074	2765	1104	2.5
003702	2443	894	2.7
003803	4132	1591	2.6
0076	3176	1256	2.5

Average number of people per household in each census tract

Two special ratios are proportions and densities. Proportions show you what part of a total each value is. For example, dividing the number of 18- to 30-year-olds in each tract by the total population of each tract gives you the proportion of people aged 18 to 30 in each tract. Proportions are often presented as percentages (the proportion multiplied by 100). Densities show the distribution of features or values per unit area. For example, by dividing the population of a county by its land area in square miles, you'd get a value for people per square mile. Density is the subject of chapter 4.

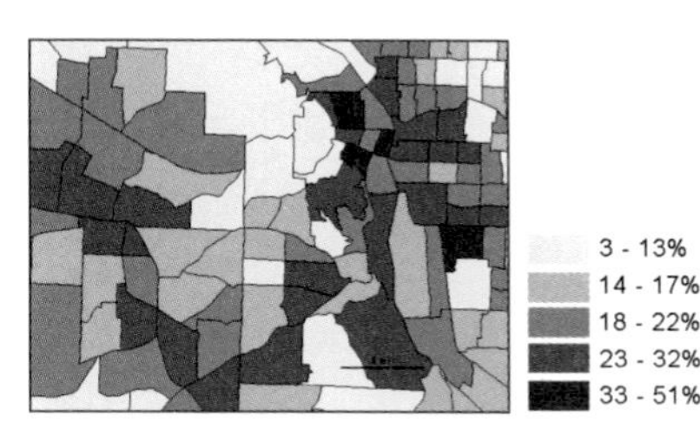

Tract	Population	18-29 Years	%18-29
003603	1606	243	15
0074	2765	516	19
003702	2443	407	17
003803	4132	751	18
0076	3176	668	21

Percentage of people aged 18 to 30 years in each census tract

Tract	Population	Square Miles	People per Sq Mi
003603	1606	0.35	4589
0074	2765	0.58	4767
003702	2443	0.37	6603
003803	4132	0.48	8608
0076	3176	0.53	5992

Population density (people per square mile) by census tract

Continuous and noncontinuous values

Categories and ranks are not continuous values—there are a set number of values in the data layer, and more than one feature may have the same value. Usually there is at least one feature having any given value. Mapping categories is discussed in chapter 2, while mapping ranks is discussed in chapter 3.

Counts, amounts, and ratios are continuous values—each feature potentially has a unique value anywhere in the range, between the highest and lowest values. That's important to realize, because knowing how the values are distributed between the highest and lowest values will help you decide how to group them for presentation, so you can see the patterns. Classifying continuous values is discussed in chapter 3.

WORKING WITH DATA TABLES

An important part of GIS analysis is working with the tables that contain the attribute values and summary statistics. Three common operations you perform on features and values within tables are selecting, calculating, and summarizing.

Selecting

You select features in order to work with a subset, or to assign a new attribute value to just those features—for example, assigning a specific rank to several different categories.

To do this, you select the rows in the data layer's attribute table that pertain to those features. You select them using a query, which is usually in the form of a logical expression:

```
select attribute = value
```

For example, to select only commercial parcels, you'd specify:

```
Select Landuse = COM
```

where Landuse is the attribute name, and COM is the value for commercial.

Parcel ID	Land Value ($)	Acres	Landuse
R916405720	10900	0.11	COM
R916405660	44400	0.28	IND
R916401590	17400	0.10	COM
R916401610	100	0.01	VAC
R710801850	42300	0.20	MFR
R710801210	230900	1.50	COM
R710801830	23600	0.08	SFR

Besides "equals," other commonly used logical operators include greater than (>), less than (<), and not equal (<>).

You can also string several expressions together to select features meeting several criteria. For example, to find commercial parcels larger than 2 acres, you'd specify:

```
Select Landuse = COM and Acres > 2
```

Parcel ID	Land Value ($)	Acres	Landuse
3S1010001500	1267080	13.81	AGR
3S1010001503	2631800	22.67	COM
3S1010001504	73640	0.46	COM
3S1010001505	420060	3.28	COM
3S1010001506	144740	0.94	COM

If you wanted to select features meeting at least one of several values or criteria, use "or." For example, to select commercial and industrial parcels, you'd specify:

```
Select Landuse = COM or Landuse = IND
```

Parcel ID	Land Value ($)	Acres	Landuse
R065301220	43700	0.11	SFR
R065301230	60000	0.16	IND
R065301250	70800	0.18	COM
R065301270	43700	0.11	SFR
R065301280	43700	0.11	SFR
R065301290	25500	0.11	COM
R065301300	68200	0.23	SFR

Calculating

You can calculate attribute values to assign new values to features in the data table. You first add a new field to the table, then assign the values for that attribute to each feature. You can assign values directly, such as with ranks, or assign values based on existing fields, such as with ratios. For example, you could select all soils of a certain type, give them a rank based on how suitable they are for growing certain crops, then do the same for other soil types in the area:

```
Select Soil = 28B
```

then,

```
Calculate Rank = 2
```

Acres	Soil Code	Rank
14	28C	
9	28B	2
153	7C	
84	29F	
22	31D	
16	35F	
15	28B	2
0	31F	

Or, you could calculate the average number of people per household in each census tract by dividing the total population by the number of households in each tract:

```
Calculate People per household = Population / Households
```

Tract	Population	Households	People per HH
003603	1606	643	2.5
0074	2765	1104	2.5
003702	2443	894	2.7
003803	4132	1591	2.6
0076	3176	1256	2.5

Summarizing

Another way of working with tables is to summarize the values for specific attributes to get statistics. In some cases you get a single value, such as a total or an average (mean); in other cases, you create a new table, listing some value for each type (category), plus a count of features. This is known as a frequency. It includes the number of features of each type and, optionally, an additional statistic, such as a total or a value.

Sum: 4896330
Count: 79
Mean: 61979
Maximum: 504000
Minimum: 0

Land value of parcels within a floodplain

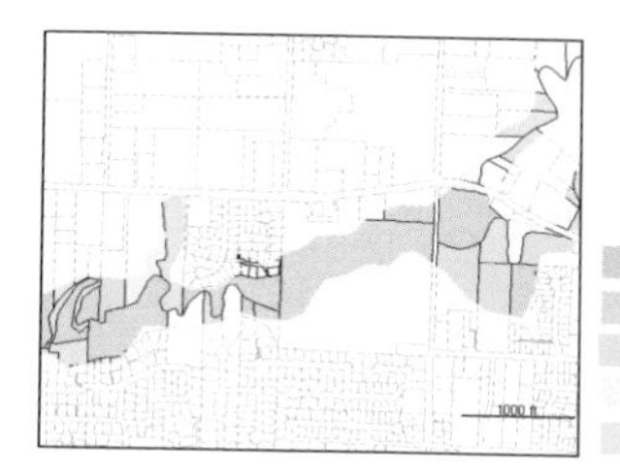

Single Family
Multi Family
Public
Rural
Agriculture
Vacant

Landuse	# of Parcels	Total SqFt
Agriculture	1	81046
Multi Family	1	137099
Public	11	1450742
Rural	6	420247
Single Family	43	788642
Vacant	20	814649

Amount of each land-use type within a floodplain

2 Mapping where things are

Mapping where things are lets you find places that have the features you're looking for, and see where to take action. You can also begin to understand why things are where they are.

In this chapter:

People often use maps to see where, or what, an individual feature is. However, by looking at the distribution of features on the map, rather than at individual features, you can see patterns that help you better understand the area you're mapping.

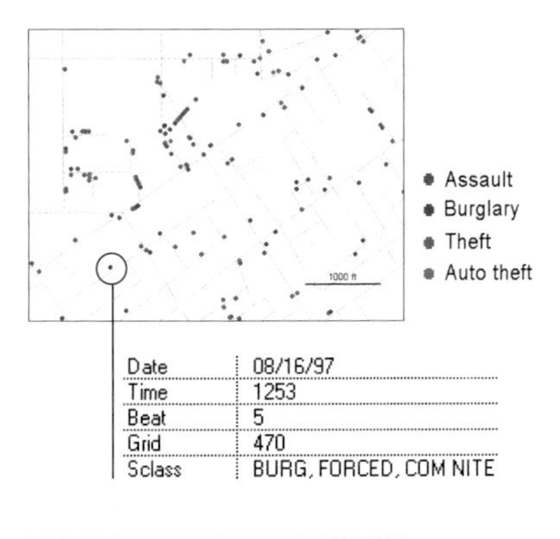

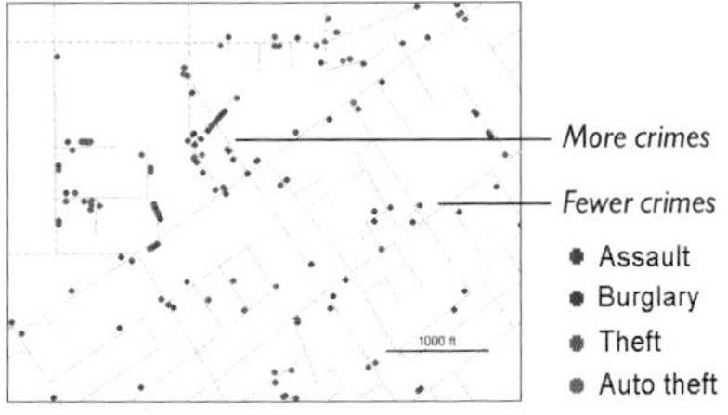

You can use a map to identify individual features or to look for patterns in the distribution of features.

Mapping where things are can show you where you need to take action, or where the areas are that meet your criteria. For example, using GIS, police can map where crimes occur each month, and whether similar crimes occur in the same place, or move to other parts of the city. This tells them where to assign patrols. Wildlife biologists studying the behavior of bears may want to find areas relatively free of roads to minimize the influence of human activity.

By looking at the locations of features, you can begin to explore causes for the patterns you see. For example, an ecologist might look at the distribution of plant communities to see if the patterns are related to terrain, rainfall, or other factors. Or, a retail analyst looking at competition between grocery stores in a region might map store locations to see how far apart stores are.

MAP GALLERY

The Oregon Department of Forestry created a map of historical land ownership, to help understand past forest management practices. Using data digitized from paper maps from various state and federal agencies, analysts grouped land parcels into 13 categories of land ownership. The map shows the large tracts of land owned by the U.S. Bureau of Land Management in the arid southeast and by the U.S. Forest Service in the west and northeast. The map is used to help manage the state's forests.

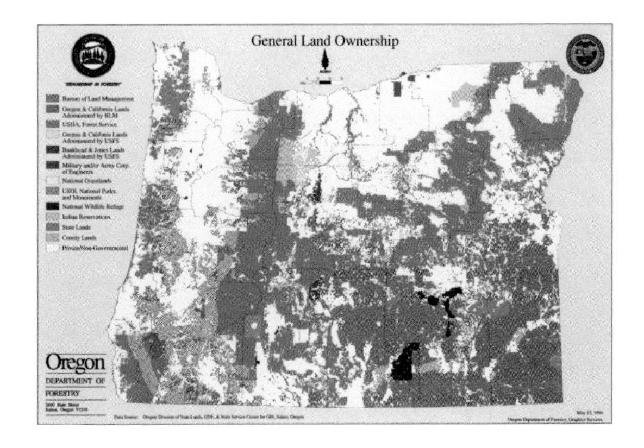

The City of North Vancouver, British Columbia, creates maps showing the condition of city streets, so repair crews can find the streets that need immediate attention. Using data from field surveys, they color code each block into one of six categories: blues and greens for better condition, oranges and reds for poorer condition. The maps show clusters of blocks needing the same type of work, which helps the city schedule crews and equipment as appropriate.

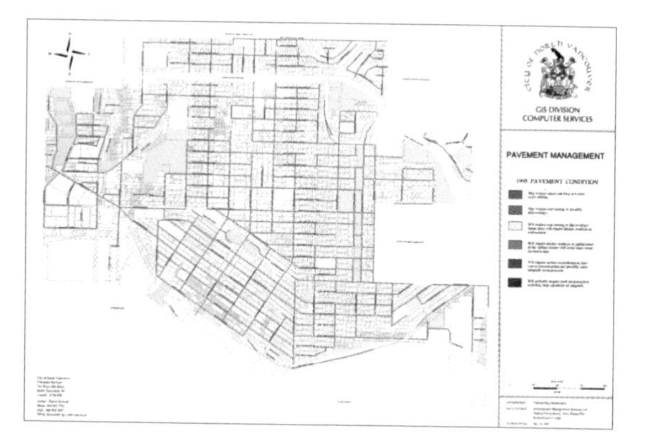

The San Diego Association of Governments (SANDAG) created a map of vegetation for the region. The map shows 15 categories of vegetation along with agricultural and developed lands. It is used in habitat conservation planning efforts to show where each habitat type is, and how much there is compared to other types.

Planners at the City of Irvine, California, mapped parcels color-coded by zoning. Similar colors are used for major categories. For example, the various commercial zoning codes are shown in shades of red. The map shows both the specific use of each parcel and the overall development pattern of the city. Planners can monitor the city's development and identify areas that will require additional services, such as water and sewer.

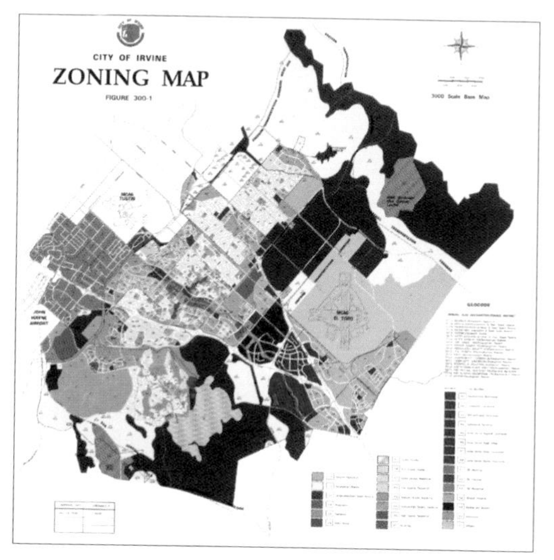

To look for geographic patterns in your data, you map the features in a layer using different kinds of symbols. You decide which features to display and how to display them based on the information you need and how the map will be used.

WHAT INFORMATION DO YOU NEED FROM THE ANALYSIS?

You might simply want to know where features are and where they're not. For example, a business might map the locations of its customers to see where to target an ad campaign. Or, a police department might map the locations of all crimes each month to see where hot spots are.

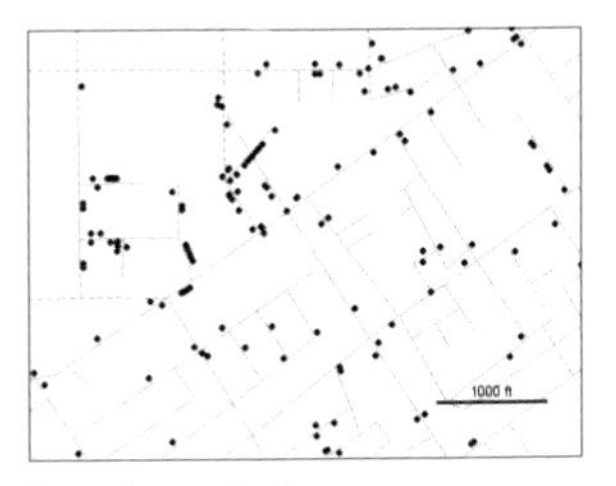

Locations of crimes

You can also use GIS to map the location of different types of features, and see whether certain types occur in the same place. A business might map customers by age category, or a police department might create a map of crimes by type—burglary, assault, theft, and so on—to see, for example, whether assaults and thefts occur in the same area.

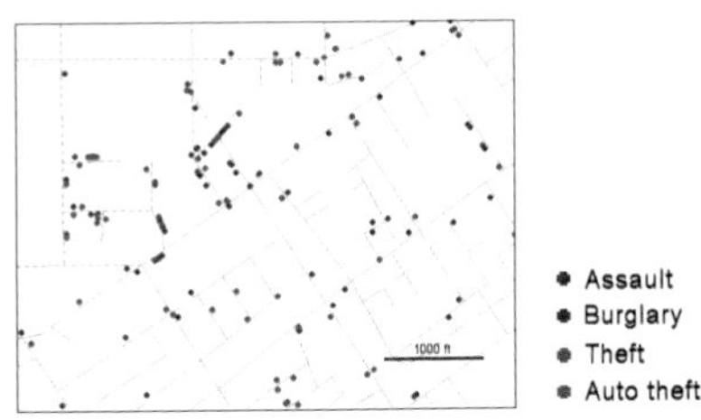

Locations of crimes, by type

HOW WILL YOU USE THE MAP?

The map should be appropriate for the audience and the issue being addressed. If, for example, a planner is showing a zoning map to the public at a city council meeting to discuss the location of heavy industry in the city in relation to high-density housing, the map would have to display detailed categories. If, on the other hand, the discussion is about overall zoning patterns in the city, displaying major categories of zoning (residential, commercial, industrial, and so on) would be sufficient. Showing the detailed categories would only add unnecessary information.

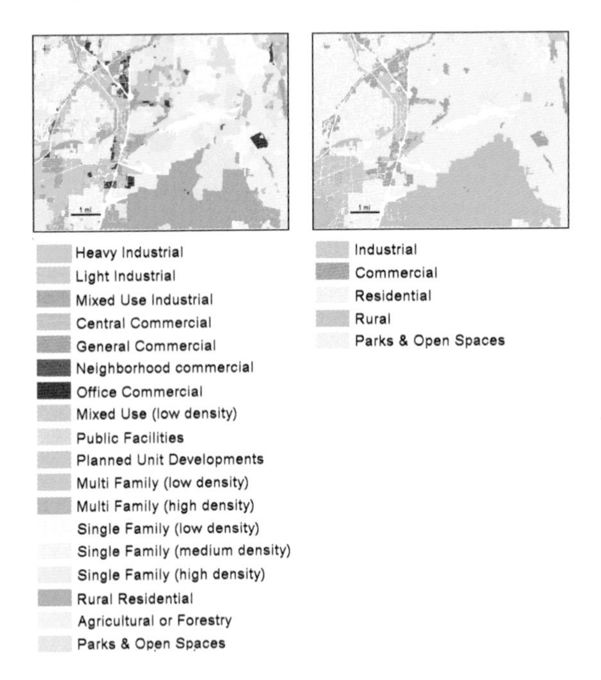

An audience that is unfamiliar with the area or the data being mapped will want to see information that provides reference locations, such as roads, lakes, or administrative boundaries.

The way in which the map will be presented also affects the amount of information you'll show. Smaller maps, such as those included in reports, should have only the information needed to show the patterns. Wall maps and posters can present more detailed data and more reference information, and still be readable.

Before creating your map, make sure the features you're mapping have geographic coordinates assigned and, optionally, have a category attribute with a value for each feature.

ASSIGNING GEOGRAPHIC COORDINATES

Each feature needs a location in geographic coordinates. If the data is already in a GIS database, the geographic coordinates have been assigned. If you're bringing the data in from another program, or entering it by hand, the features will need to have location information such as a street address, or latitude–longitude values. The GIS reads these and assigns geographic coordinates.

ASSIGNING CATEGORY VALUES

When you map features by type, each feature must have a code that identifies its type, for example, whether a crime is an assault, theft, burglary, and so on. This information may already be stored with each feature, or you may need to add it. To add a category, you create a new attribute in the layer's data table and assign the appropriate value to each feature.

Many categories are hierarchical, with major types divided into subtypes. For example, parcels may have a general code indicating that they're zoned industrial, and a detailed code indicating the type of industrial zoning: heavy, light, or mixed use.

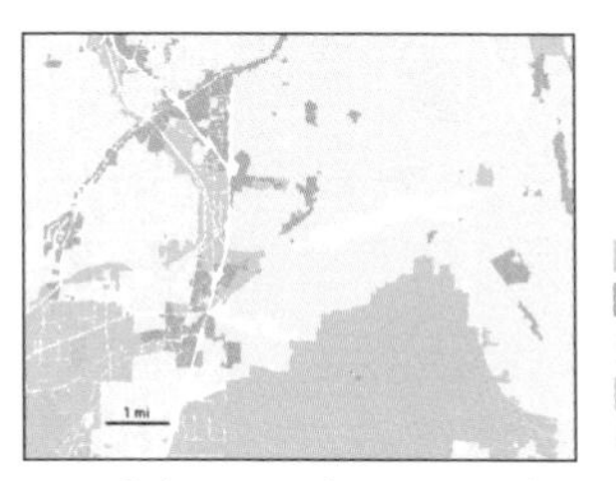

Parcels by general zoning code

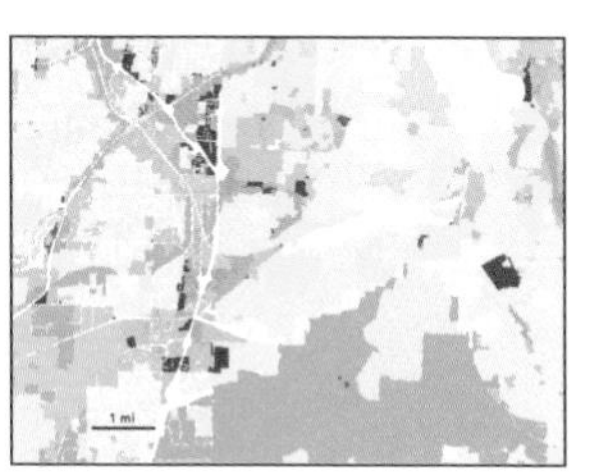

Parcels by detailed zoning code

In some cases, a single code indicates both the major type and subtype. For example, all crimes with a value between 500 and 599 are burglaries, but the type of burglary is indicated by the specific value.

ID	Date	Type	Description
108161454	08/11/97	629	THEFT, $200-$400
107941626	07/20/97	521	BURG, UNL ENT, RES NITE
109040815	11/07/97	521.1	BURG, UNL ENT, GAR NITE
106270910	02/03/97	513	BURG, FORCED, RES UNK

In other cases, the separate attributes are used to store the major types and subtypes, as in this zoning layer.

Acres	General Code	Detailed Code	Description
3.8	SFR	SFR2	Urban Low Density Residential (10000 sq. ft.)
17.6	COM	CO	Office Commercial
4.7	RUR	RRFU	Future Urban
104.9	SFR	SFR1	Urban Low Density Residential (30000 sq. ft.)
2.9	SFR	SFR3	Residential - 7500 sq. ft. area per unit
46.9	MFR	MFR1	Medium Density Residential
8.4	SFR	SFR2	Low Density Residential
7776.2	RUR	RRFU	Rural Residential/Farm Forest (5 Acres)
2650.0	RUR	RRFU	Rural Residential/Farm Forest (5 Acres)
4681.6	RUR	FF	Exclusive Farm Use

To create your map, you tell the GIS which features you want to display, and what symbols to use to draw them. You can map all features in a layer as a single type, or show them by category values.

MAPPING A SINGLE TYPE

To map features as a single type, you draw all features using the same symbol. Even these very basic maps—which simply show where features are—can reveal patterns. For example, a store owner can see where customers are, or a recreation planner can see where roads in a natural area are.

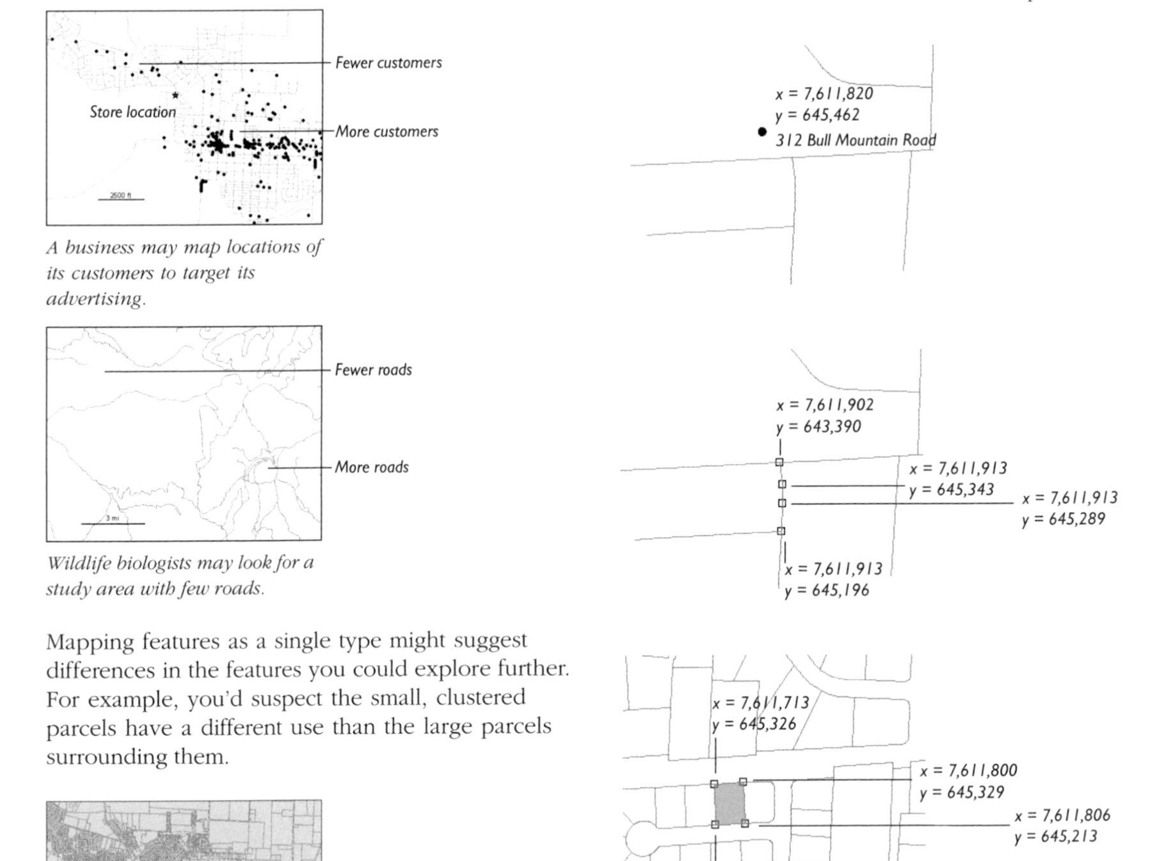

A business may map locations of its customers to target its advertising.

Wildlife biologists may look for a study area with few roads.

Mapping features as a single type might suggest differences in the features you could explore further. For example, you'd suspect the small, clustered parcels have a different use than the large parcels surrounding them.

What the GIS does

The GIS stores the location of each feature as a pair of geographic coordinates, or as a set of coordinate pairs that define its shape (line or area). When you make a map, the GIS uses the coordinates to draw the features, using a symbol you specify. For individual locations, such as customer addresses, the GIS draws a symbol at the point defined by the coordinates for each address. For linear features, such as streets, the GIS draws lines to connect the points that define the shape of each street. For areas, such as parcels of land, the GIS draws their outlines or fills them in with a color or pattern.

Coordinate pairs define the location of an address, the shape of a street, and the boundary of a parcel.

Using a subset of features

You can map all features in a data layer or a subset you've selected based on a category value. For example, you could map all crimes, select all burglaries and map those, or select and map only commercial burglaries. Using a subset can reveal patterns that aren't apparent when mapping all features.

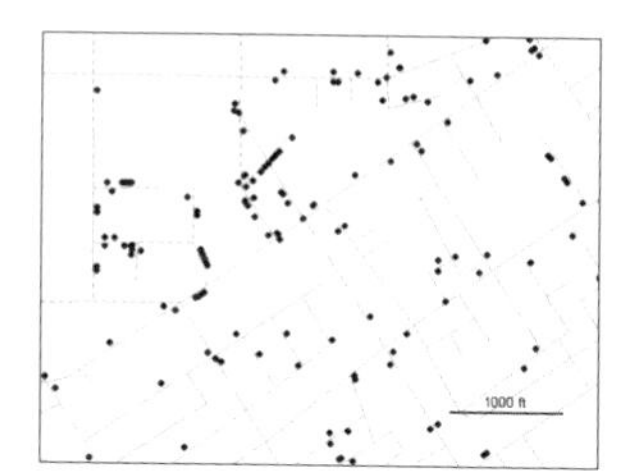

All crimes

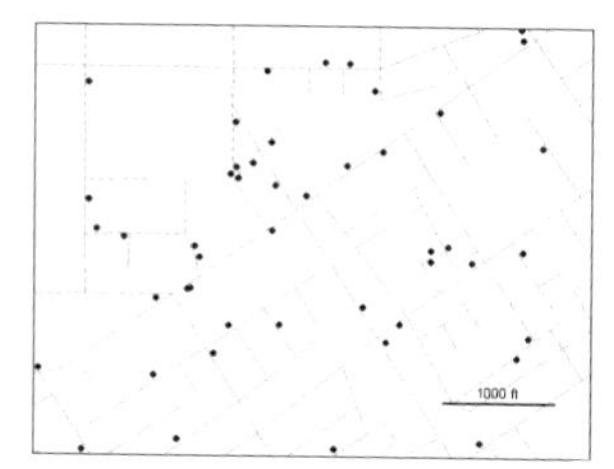

Burglaries

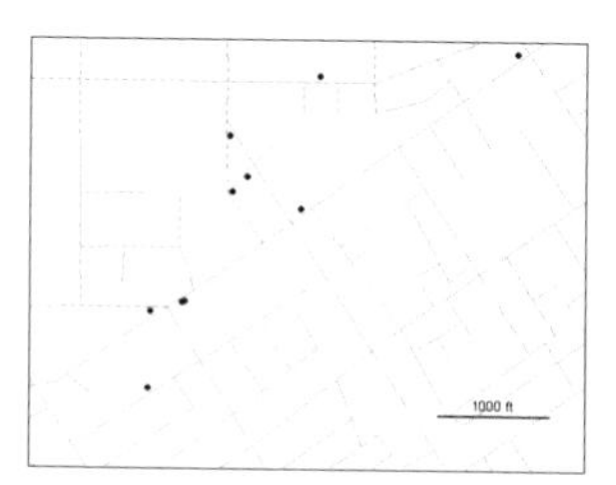

Commercial burglaries

Mapping a subset is more commonly done for individual locations. Because linear features are often connected in a network, a map showing a subset of the features could be incomplete. For example, a map showing only local roads without freeways and highways would not show all the connections in the road network.

Similarly, showing a subset of continuous data leaves the features without a context. For example, by showing only areas zoned for commercial use, you wouldn't know what type of zoning surrounds those areas. If you wanted to highlight commercial zoning, you'd show the other types as well, but draw them in a lighter shade.

Commercial zoning

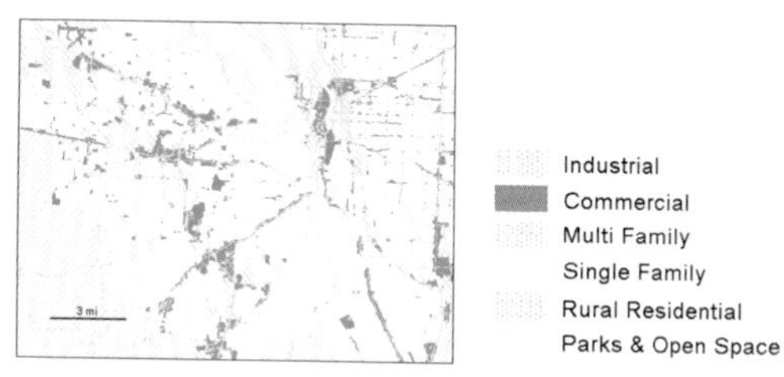

All zoning, with commercial highlighted

MAPPING BY CATEGORY

You can map features by category, by drawing features using a different symbol for each category value. Mapping features by category can provide an understanding of how a place functions. For example, mapping all major roads using a black line merely shows you the locations of the roads. Mapping them by the type of road shows you the hierarchy of roads, and in turn, the regional traffic patterns; you can see that arterials and highways feed traffic into the central loop freeways.

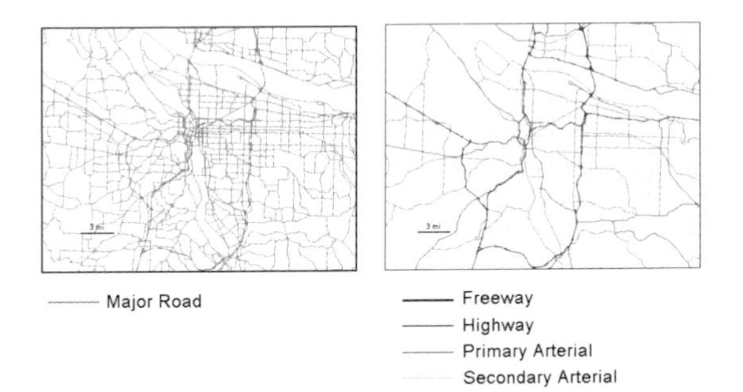

Mapping major roads by category gives a better sense of regional traffic patterns than simply mapping all roads using the same symbol.

Similarly, mapping crimes by type shows you which types of crimes occur where. It also shows which types occur near each other, in a given time and place.

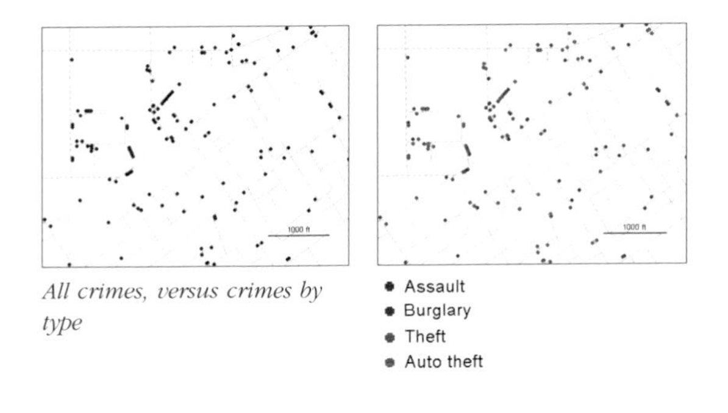

All crimes, versus crimes by type

What the GIS does

The GIS stores a category value for each feature in the layer's data table. It also stores, separately, the characteristics of the symbols you specified to draw each value. When you display the features, the GIS looks up the symbol for each feature based on its category value, and uses that symbol to draw the features on the map. For example, you could have the GIS draw four-lane streets with a thick line and two-lane streets with a thin one.

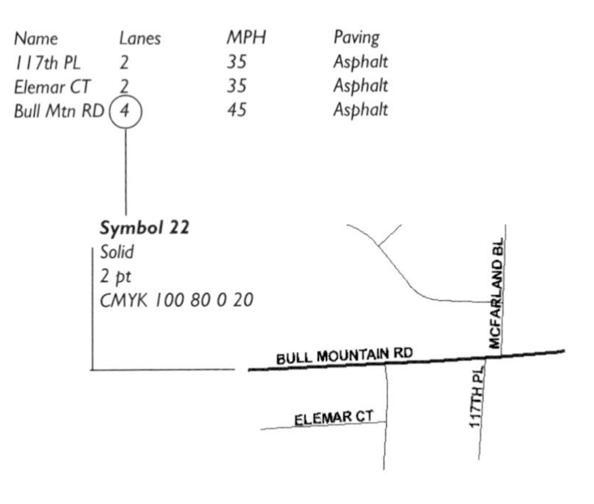

Displaying features by type

Features might belong to more than one category. Using different categories can reveal different patterns. For example, you could show burglaries by the type of building entered (residential or commercial), or by the type of entry (forced, or not).

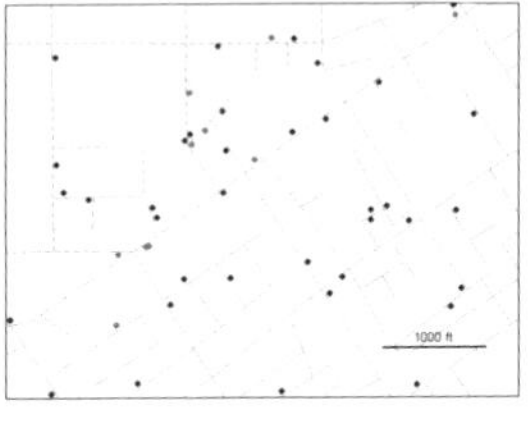

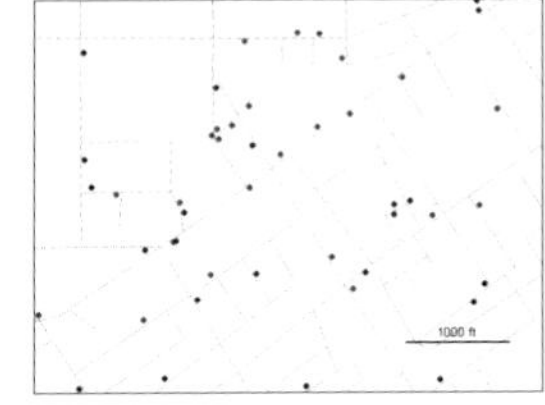

- Commercial Burglary
- Residential Burglary

- Forced Entry
- Non-Forced

Usually, several categories are shown on the same map. However, if the patterns are complex or the features are close together, creating a separate map for each category can make patterns within a particular category—and even across categories—easier to see. This is especially true when mapping individual locations, such as businesses, or crimes.

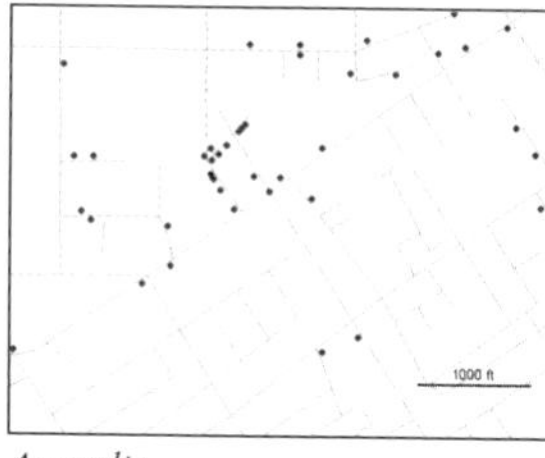

Assaults

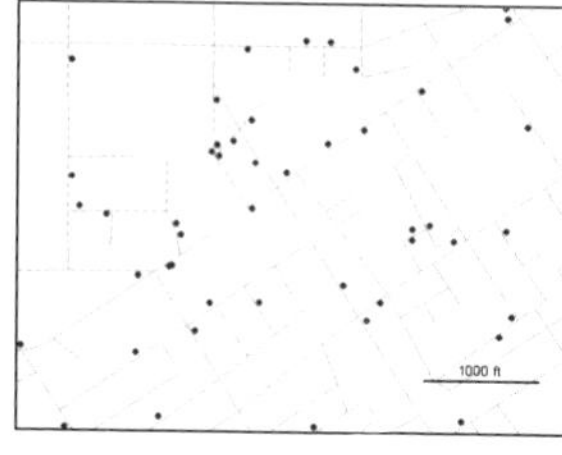

Burglaries

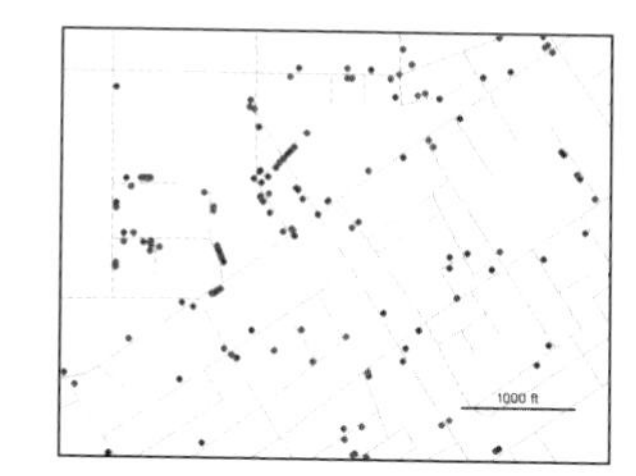

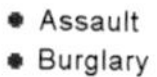

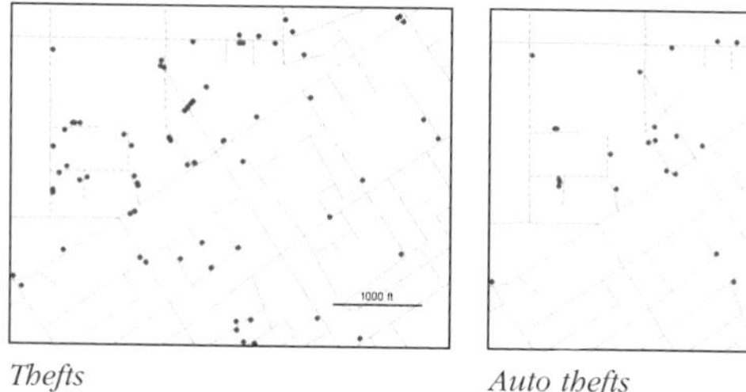

Thefts

Auto thefts

Displaying a subset of categories may make it easier to see if different categories are related. For example, it's easier to see that commercial and multifamily land use occur near each other in this region, when you display only those two categories.

How many categories

If you're showing several categories on a single map, you'll want to display six or seven categories at most. Because most people can distinguish up to seven colors or patterns on a map, displaying more categories than this makes the patterns difficult to see. The distribution of features and the scale of the map also affect the number of categories you can display.

The features being mapped

If the map contains small scattered features rather than large contiguous ones, readers will find it difficult to distinguish the various categories. If the features are sparsely distributed, you can display more categories than if the features are dense.

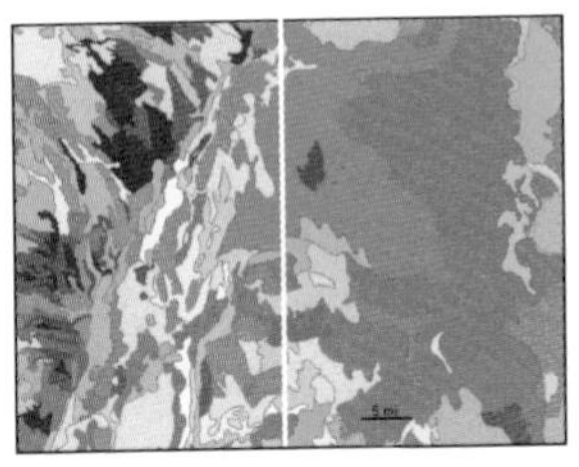

The left half of this vegetation map has many small areas, compared to the right half; so it's hard to distinguish the various categories, and thus the patterns, in the map.

Map scale

When mapping an area that's large relative to the size of the features, using more than seven categories can make the patterns difficult to see. In the top set of zoning maps, the patterns are clearer when fewer categories are used.

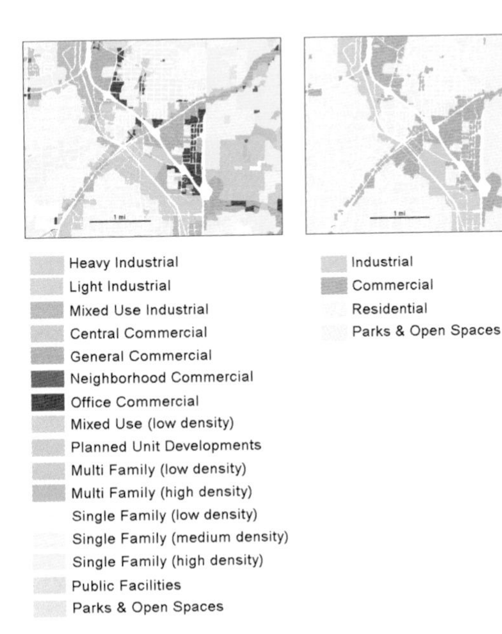

When smaller areas are mapped, individual features are easier to distinguish, so more categories will also be easier to distinguish. In fact, using too few categories can cause important information to be left out. The map on the right below shows the overall patterns of industrial, commercial, and residential zoning. However, it doesn't distinguish between the various types of commercial and residential zoning, information that may be important for a developer or planner. The map on the left shows this information.

Grouping categories

If you have more than six or seven categories, you may want to group them to make the patterns easier to see. The same set of data can be used to generate different maps. The map on the left below uses eighteen zoning categories. With this many categories, it's hard to distinguish between some of them, such as "Central Commercial" and "Planned Unit Developments." In the second map, the eighteen categories have been grouped into five categories and the zoning pattern for the area is apparent at a glance. The trade-off is that some important information may be lost (the location of parcels zoned for offices, for instance). Using fewer categories makes understanding easier for a broader audience, but reduces the amount of detailed information the map presents.

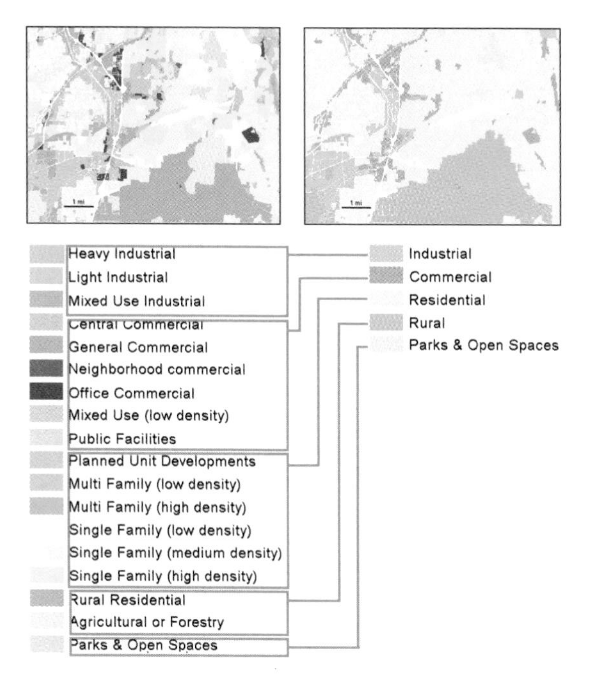

Patterns may be easier to see if you generalize many categories into a few.

The way in which you group the categories can change the way readers perceive the information. Simply reassigning a class of features from one general category to another can create two quite different maps. In the map on the left, using the same zoning data as before, "Rural Residential" has been coded as "Residential" instead of "Rural," and "Agricultural or Forestry" has become a separate category. Compare this map with the map on the right, in which "Rural Residential" is combined with "Agriculture or Forestry." The distinct edge between urban and rural has disappeared; the new map shows an urbanized area with small patches of agriculture, forestry, and parks.

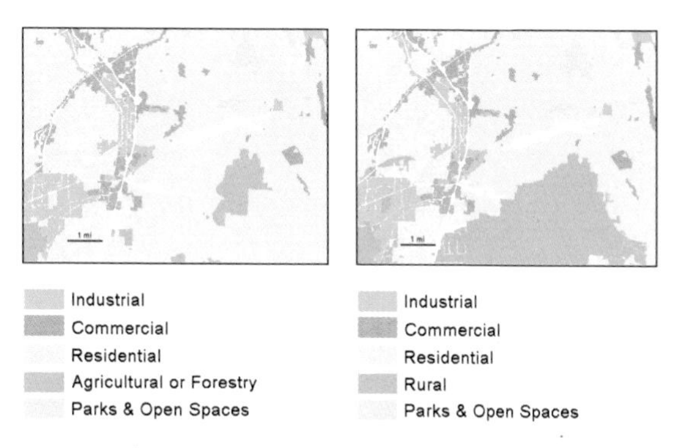

Assigning features to categories in various ways can affect the patterns on the map.

To decide how to group and display data, you have to understand what it represents. For example, "Rural Residential" includes areas zoned for future urban development. To show current conditions, you'd group "Rural Residential" with agriculture and forestry into a "Rural" category. But if you want to show what the future development patterns might look like, you'd group it with "Residential" or another urban category. It's important to be explicit (either on the map or in a report) about what the categories include so map readers know what the map represents.

You can group categories in several ways. One way is to assign each record in the database two codes: one for its detailed category and one for its general category. You can do this fairly easily by selecting all the features with the appropriate detailed code value, and assigning the general code value to all of them at once. The table shown here contains a few records from the zoning database showing both the detailed code and the general code. Notice the various residential types all have the same general code, "RES." With this method, all the codes are stored together in one table, simplifying the database. However, if you change the category groupings, you'll have to go through the selection and assignment process again.

Description	Detailed Code	General Code
Urban Low Density Residential (30000 sq. ft.)	SFR1	RES
Residential - 7500 sq. ft. area per unit	SFR3	RES
Medium Density Residential	MFR1	RES
Low Density Residential	SFR2	RES
Rural Residential/Farm Forest (5 Acres)	RRFU	RUR
Rural Residential/Farm Forest (5 Acres)	RRFU	RUR
Exclusive Farm Use	FF	RUR

Option 1: Assign a general code to each record in the database.

Another way to group categories is to create a table containing one record for each detailed code, with its corresponding general code. When you're ready to display the map, you join the feature database table with the new table, and use the general code to display the features. This requires some data entry, but it's easy to change the category groupings. Any new features you add are automatically assigned a general code when you rejoin the tables.

Description	Detailed Code
Urban Low Density Residential (30000 sq. ft.)	SFR1
Residential - 7500 sq. ft. area per unit	SFR3
Medium Density Residential	MFR1
Low Density Residential	SFR2
Rural Residential/Farm Forest (5 Acres)	RRFU
Rural Residential/Farm Forest (5 Acres)	RRFU
Exclusive Farm Use	FF

Detailed Code	General Code
FF	RUR
RRFU	RUR
SFR1	RES
SFR2	RES
SFR3	RES

Option 2: Create a linked table to match detailed codes with general codes.

A third method of grouping categories is to simply assign the same symbol to the various detailed categories that comprise each general category, when you make the map. The GIS lets you save the symbol assignments in a file and reuse them if you need to recreate the map. This method avoids having to modify the database or create new data tables, but the general codes aren't available in the database for other analysis you may want to do.

	Codes	Category
	CC, CG, CN, CO	Commercial
	FF	Agricultural or Forest
	IH, IL, IMU	Industrial
	MFR1, MFR2	Multi Family

Option 3: Assign categories on the fly by specifying symbols.

Choosing colors and symbols

The colors and symbols you use to display the categories can help reveal patterns in the data.

- If you're mapping individual locations, you can use a single symbol in a different color for each category, or a different symbol shape for each category, or both. Using symbols alone can make the patterns harder to see because, for point symbols, it's harder to distinguish shapes than colors.

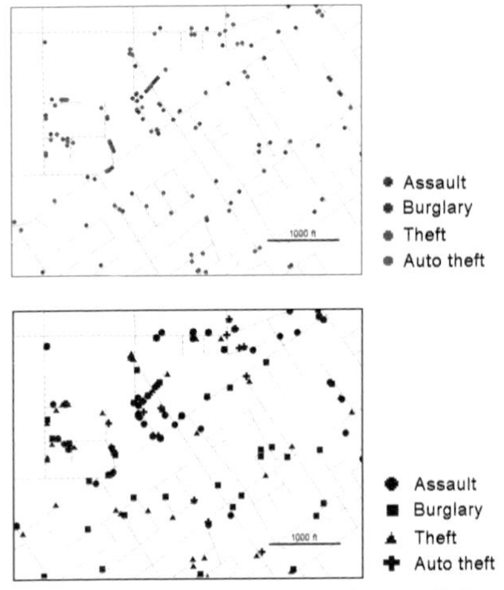

Colors are easier to distinguish than symbols.

- If you'll be printing the map, test to make sure point symbols are large enough to be seen at the scale you're using, but not so big that they obscure other features. Maps displayed on screen—such as on the Internet—should use larger, simpler symbols than printed maps, as printers have better resolution than screen displays.

- If you're mapping linear features, use different widths or symbols (such as double or dashed lines) to distinguish the categories. Since most linear features are narrow, color alone is not enough to let readers distinguish the categories. If you make the lines wide enough, though, you can also use color to distinguish the category. Different widths and symbols are often used if there is an implied ranking in the category—for example, freeways are usually drawn wider than highways, which are in turn drawn wider than local streets. Combining colors and symbols can help distinguish categories. Highway maps often combine line width and color to distinguish the type of road.

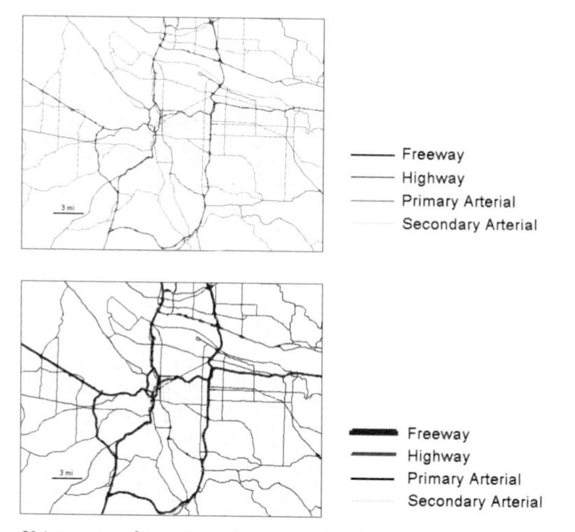

Using a combination of color and pattern can make the patterns easier to see.

- For shaded areas or raster layers, displaying similar categories in different shades of the same color can make overall patterns clearer, but may also make these categories difficult to distinguish.

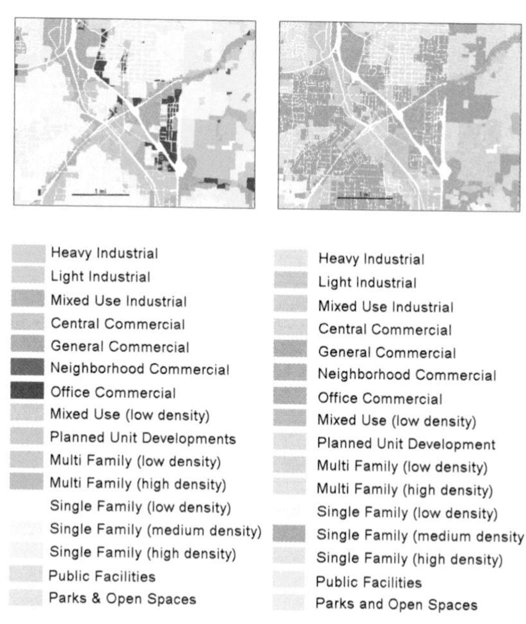

Using similar colors for related categories—rather than randomly assigned colors—can help reveal patterns.

- Text labels can also help distinguish categories. Soil and geology maps often use a two- or three-letter code to label each feature on the map, and include the full name of the soil or geologic type in the legend.

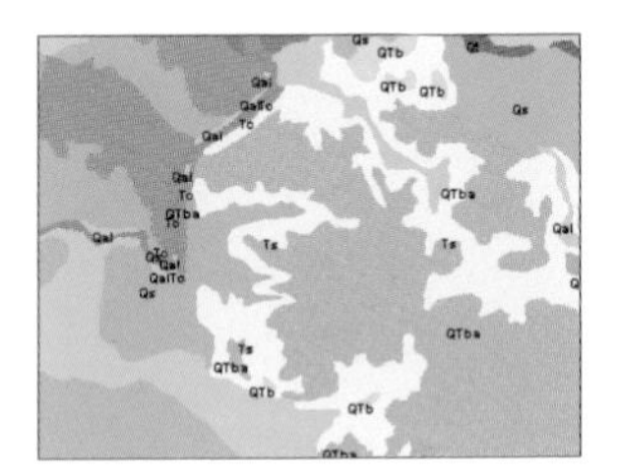

OW	Open Water
QTb	Basalt
QTba	Basalt and Basaltic Andesite
QTs	Sedimentary Rocks
Qal	Alluvial Deposits
Qgs	Glacial Deposits
Qls	Landslide and Debris Flow Deposits
Qs	Lacustrian and Fluvial Sedimentary Rocks
Qt	Terrace, Pediment, and Lag Gravels
Tbaa	Andesitic and Basaltic Rocks
Tc	Columbia River Basalt Group
Tcg	Clastic Rocks
Tfc	Fanglomerate
Ts	Tuffaceous Sedimentary Rocks
Tub	Basaltic Lava Flows
Tus	Sedimentary and Volcaniclastic Rocks

Mapping reference features

Your map will be more meaningful for people if you display recognizable landmarks such as major roads or highways, administrative or political boundaries, locations of towns or cities, or major rivers.

You may also want to map reference features specific to your analysis, so you can look at geographic relationships. For example, if you're mapping customers, you'll also want to map the locations of your stores.

Map reference features using light grays, or other light colors such as pale greens and blues, so that they don't dominate the map. If you're mapping individual locations (such as schools, stores, or ranger stations) as reference features, you may need to draw them using a large symbol or more noticeable color to make them more visible.

If your map presents the information clearly, you may be able to see some patterns in the data. If you're mapping a single category, you may see that features are clustered, uniformly spaced, or randomly distributed. The patterns are partly dependent on the scale of the map. By zooming in or out, you may see patterns that were not evident before.

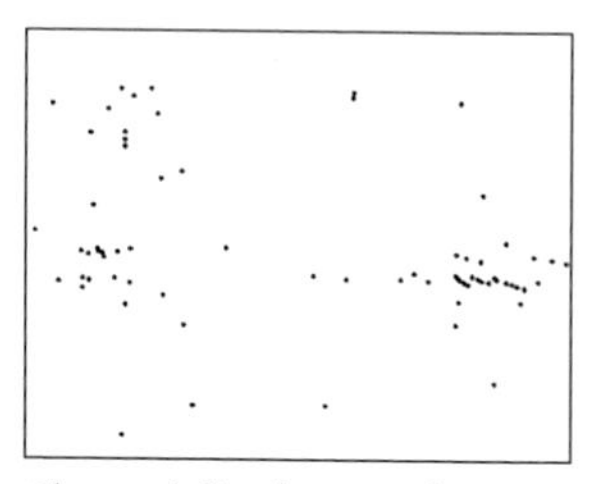

Clustered distribution—features are more likely to be found near other features.

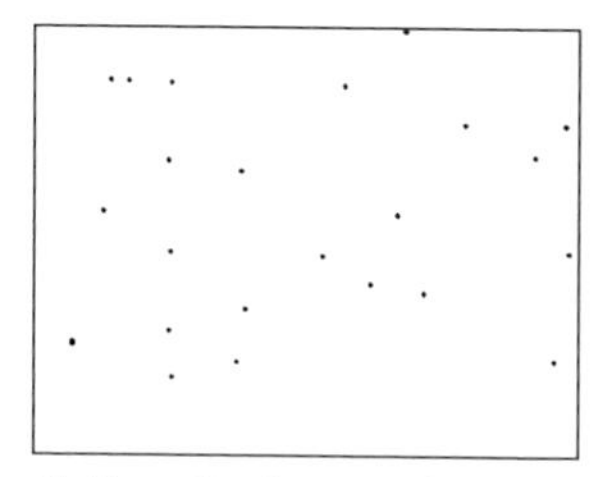

Uniform distribution—features are less likely to be found near other features.

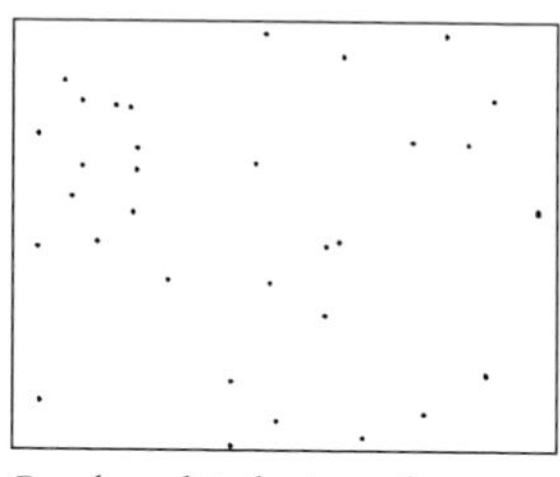

Random distribution—features are equally likely to be found at any given location.

Features in a particular category might be clustered, or might be found with other features. In the map below, retail, financial, and service businesses cluster along major streets and intersections. This information would be useful for a salesperson visiting these types of businesses.

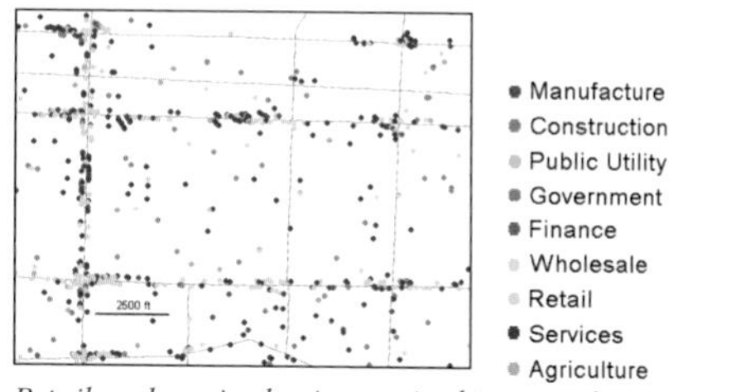

Retail and service businesses in this area cluster at major intersections and along major streets.

Based on what you know about a place, or similar places, you may discover that the patterns are meaningful. This map of parcels shows the development pattern common to many towns: a central commercial strip abutted by manufacturing and multifamily housing, with single-family housing around the edges, and beyond that, agriculture. You can begin to understand how the town developed, and where it might grow.

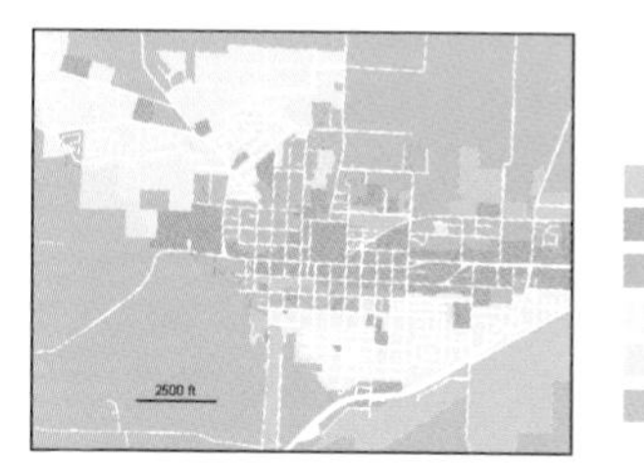

This development pattern is typical of many small towns.

The patterns on a map might suggest a reason for why things are where they are. In the map on the left, the pattern of north–south bands of vegetation types is clear. Since elevation affects where different kinds of plants grow, you might start by looking at the local terrain (right map) to see if it echoes this north–south pattern.

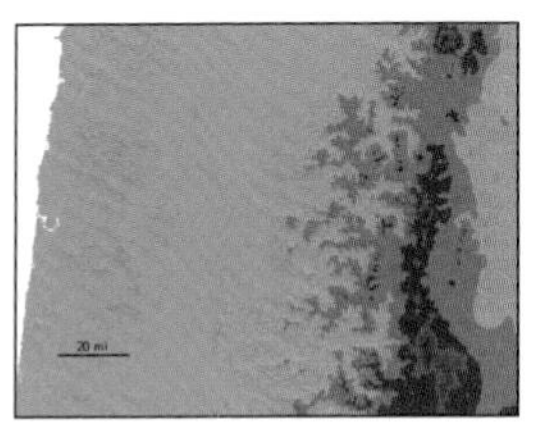

Topography is one likely cause of the strong north–south lines of vegetation.

A pattern can be the result of several factors. On the map below, each dry cleaner is a little over a mile from the next (with the exception of the two shops in the lower center of the map), suggesting that the market area for each is about a half-mile radius around the shop. The dry cleaners are also all located on major arterial streets. To find out if other factors are involved, you'd have to look at population density in this area, the characteristics of each shop, and other factors.

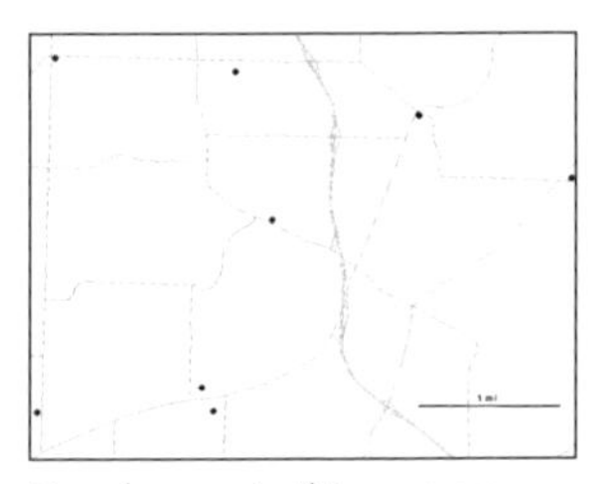

Dry cleaners in this area are located about a mile apart along major streets.

Exceptions, such as the two dry cleaners located near each other, can reveal further causes. Either there are enough customers to support both of them—perhaps there is an office complex nearby—or they offer different services.

You can see many patterns just by looking at a map. However, to find out if there are hidden patterns in the data or whether the patterns you see are meaningful, you need to use statistics to measure and quantify the relationships between features. You might do this if you are conducting scientific research, or if there could be legal implications from the analysis. A number of standard statistical methods for geographic analysis can be found in the references listed in the section 'Where to get more information.'

Mapping the most and least

Mapping the most and least lets you compare places based on quantities so you can see which places meet your criteria, or understand the relationships between places.

In this chapter:

- *Why map the most and least?*
- *What do you need to map?*
- *Understanding quantities*
- *Creating classes*
- *Making your map*
- *Looking for patterns*

People map where the most and least are to find places that meet their criteria and take action, or to see the relationships between places. To map the most and least, you map features based on a quantity associated with each. For example, a catalog company selling children's clothes would want to find ZIP Codes with many young families with relatively high income. Or, public health officials might map the numbers of physicians per 1,000 people in each census tract to see which areas are adequately served, and which are not.

Mapping features based on quantities adds an additional level of information beyond simply mapping the locations of features. For example, mapping the locations of businesses gives a sense of where workers are, information that might be useful for a transportation planner. But mapping the businesses based on the number of employees at each business gives a much better picture of where employees are.

Locations of businesses

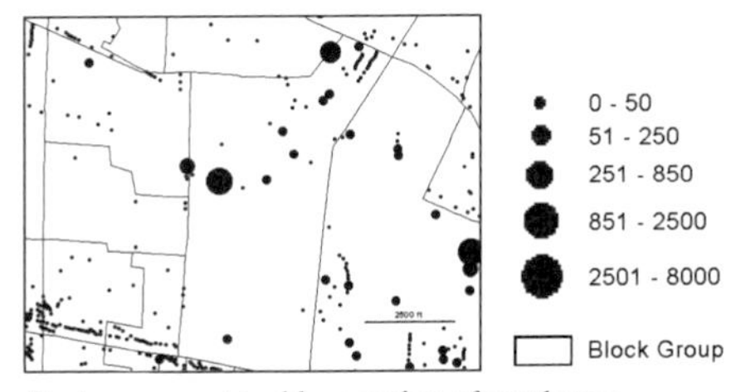

Businesses mapped by number of employees

MAP GALLERY

Police in Stockton, California, mapped locations where truant students were picked up. The size of the red dot indicates the number of children picked up at that location. The map lets school officials and police see where the highest truancy is, so they can focus on those areas.

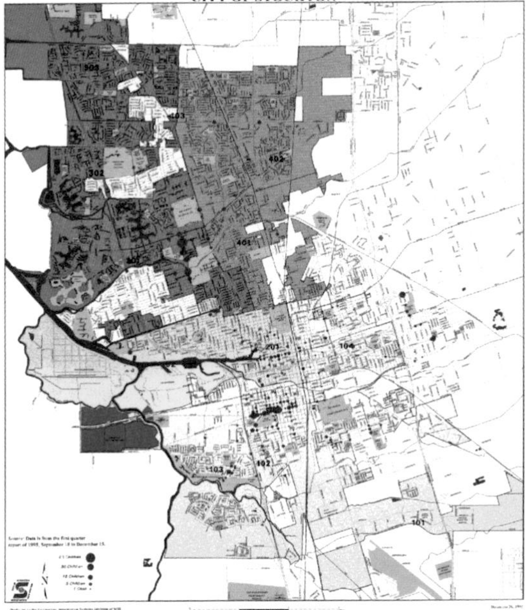

Analysts at the Department of Public Works in Louisville, Kentucky, created a map showing the location of census block groups qualifying for the Community Development Block Grant program. In the block groups shaded green, at least 51 percent of the people are of low and moderate income. The map also shows the city's ward boundaries, so aldermen can see what portions of their wards qualify for the grant program.

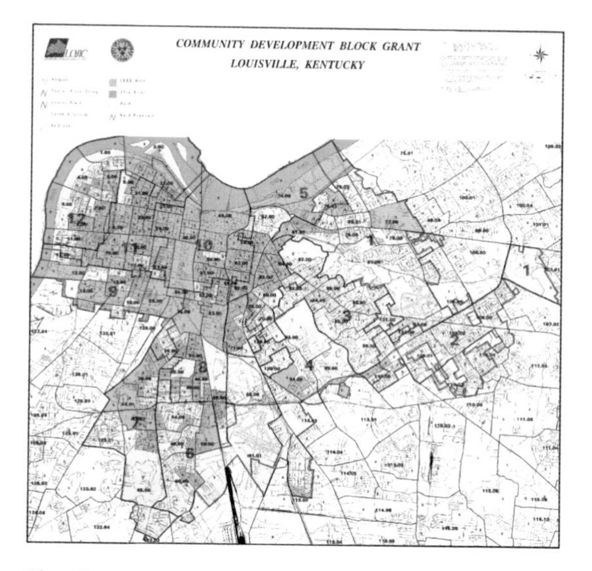

The Qatar Ministry of Public Health mapped public health statistics to compare the quality of health care across the country. Using records from state-run health care centers, they mapped the number of visits per person, the percentage of visits in which prescriptions were given, the percentage in which lab referrals were given, and other statistics. The maps were created by shading the areas covered by the health care centers, based on the statistics. Some differences among the centers can be seen. For instance, the centers in the south tend to prescribe drugs more often (indicated by the dark orange) and refer cases to labs less often (indicated by the light yellow). Officials use the maps to see how and where health care varies.

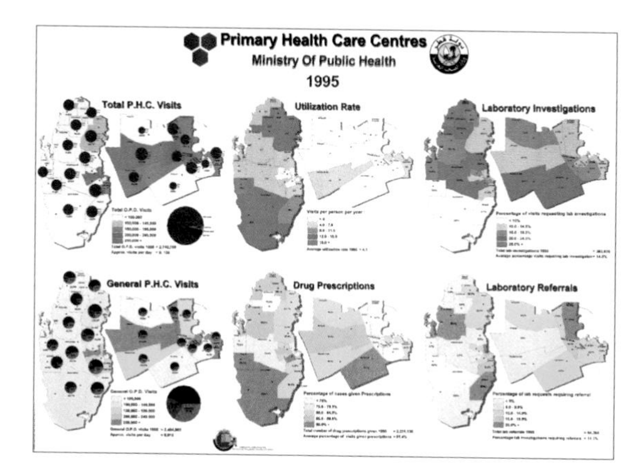

Analysts at Southern California Edison mapped the number of calls for service for a three-month period to see how different parts of the service area compared. They drew the location of each electric meter color-coded by the number of calls for service, from light red for few calls to dark red for many. The map shows that fewer calls were made in the southern portion of the area. There are several small areas where many calls were made, indicating a power outage or other problem affecting a number of people. The company uses the map to help find out why certain areas have more calls and others fewer.

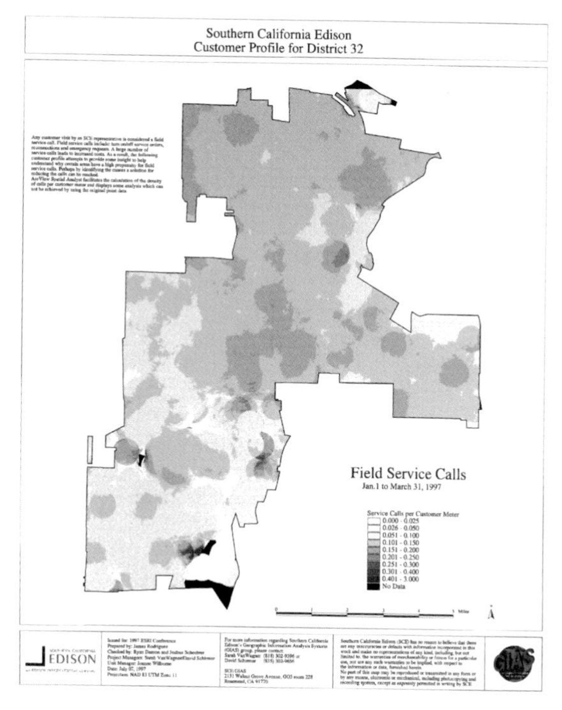

By mapping the patterns of features with similar values, you'll see where the most and least are. Knowing the type of features you're mapping, as well as the purpose of your map, will help you decide how to best present the quantities to see the patterns on your map.

WHAT TYPE OF FEATURES ARE YOU MAPPING?

You can map quantities associated with discrete features, continuous phenomena, or data summarized by area.

Discrete features can be individual locations, linear features, or areas. Locations and linear features are usually represented with graduated symbols, while areas are often shaded to represent quantities.

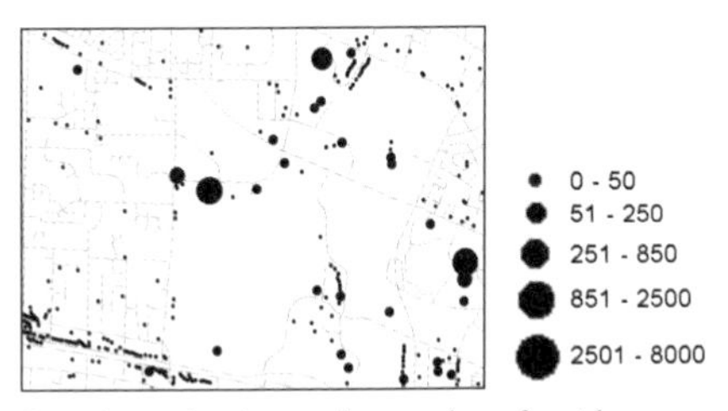

Locations—businesses by number of employees

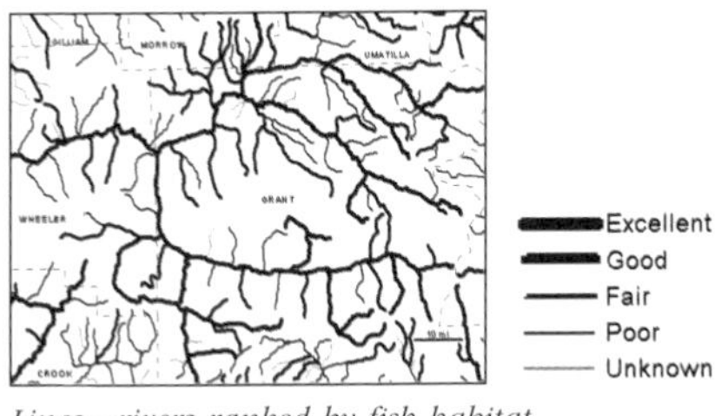

Lines—rivers ranked by fish habitat

Areas—parcels by land value ($US per square foot)

Continuous phenomena can be defined areas or a surface of continuous values. Areas are displayed using graduated colors; surfaces are displayed using graduated colors, contours, or a 3-D perspective view.

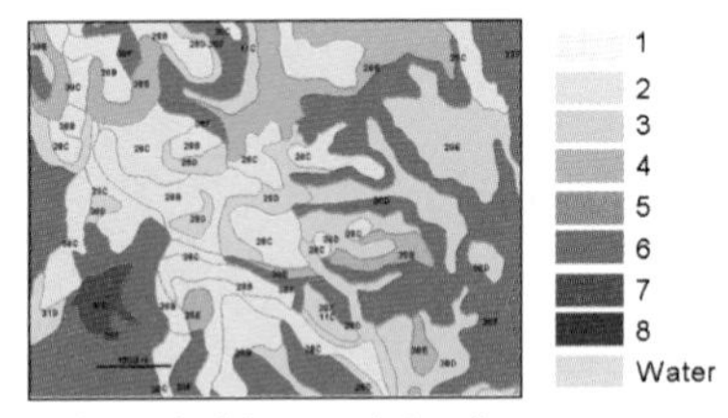

Soils ranked by suitability for growing crops. Soils with a rank of 1 are the most suitable.

A surface of land value ($US per square foot) created by interpolating from tax parcel centroids. The lighter areas have a higher value.

Data summarized by area is usually displayed by shading each area based on its value, or using charts to show the amount of each category in each area. You can summarize individual locations, linear features, or areas. Chapter 5, 'Finding what's inside,' discusses summarizing by area.

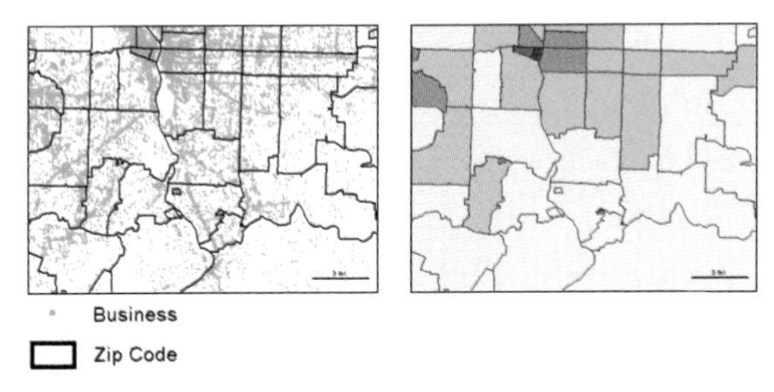

The number of businesses in each ZIP Code is summarized to create a map showing ZIP Codes shaded by total number of businesses per square mile. The darker shades indicate a higher concentration of businesses.

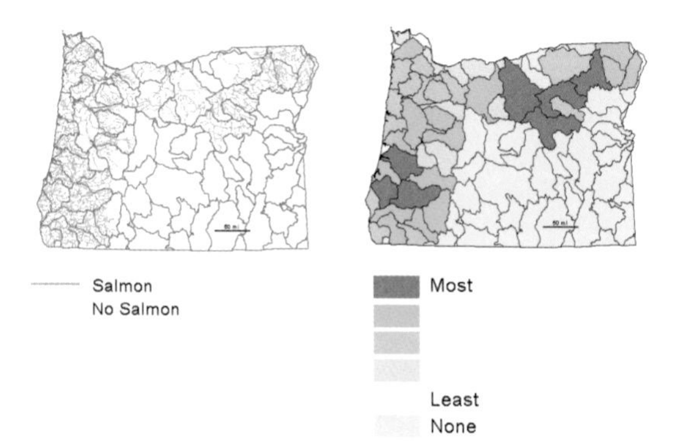

Streams summarized by watershed show which watersheds have the longest salmon runs.

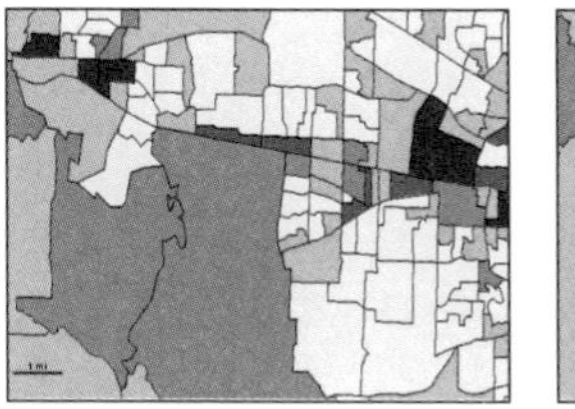

0 - 5
5.1 - 10
10.1 - 15
15.1 - 20
> 20

Poverty rate by block group (left) and by census tract

ARE YOU EXPLORING THE DATA OR PRESENTING A MAP?

Keeping the purpose of your map and the intended audience in mind will help you decide how to present the information on your map.

You might be exploring the data to see what patterns and relationships you can find. For example, you might look at the distribution of median income in an area to see the range of values, and where high- and low-income families cluster. When you're exploring, you'll display the data in more detail and try various ways of displaying it.

In another case, you might want to present a map showing specific patterns that answer a particular question. For example, you may want to show which areas have at least 35 percent of families living below poverty level, and which areas have 50 percent or more, so city officials can see areas potentially qualifying for economic development grants. When creating a map for presentation, you'll generalize the data to reveal the patterns.

In many cases, you'll start by exploring the data to see what patterns emerge, and what questions arise, and later create a generalized map to reveal specific patterns.

To map the most and least you assign symbols to features based on an attribute that contains a quantity. Quantities can be counts or amounts, ratios, or ranks. Knowing the type of quantities you're mapping will help you decide the best way to present the data.

Tract	Population	18-29 Years	%18-29
003603	1606	243	15
0074	2765	516	19
003702	2443	407	17
003803	4132	751	18
0076	3176	668	21

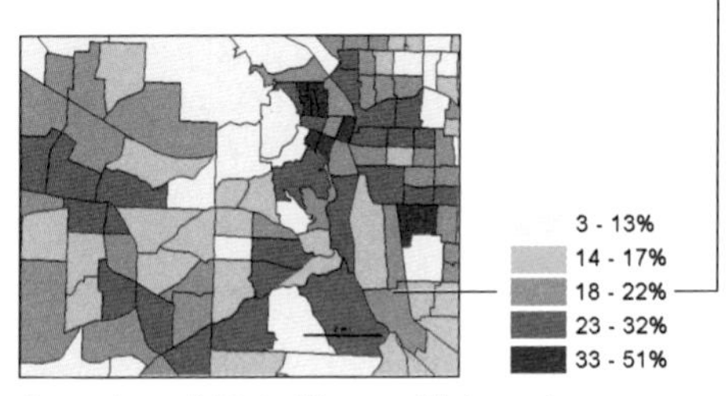

Percentage of 18- to 29-year-olds in each census tract

COUNTS AND AMOUNTS

Counts and amounts show you total numbers. A count is the actual number of features on the map. An amount is the total of a value associated with each feature. Using a count or an amount lets you see the value of each feature as well as its magnitude compared to other features.

You can map counts and amounts for discrete features, for example, the number of employees at each business; or for continuous phenomena, for example, the annual precipitation at any location.

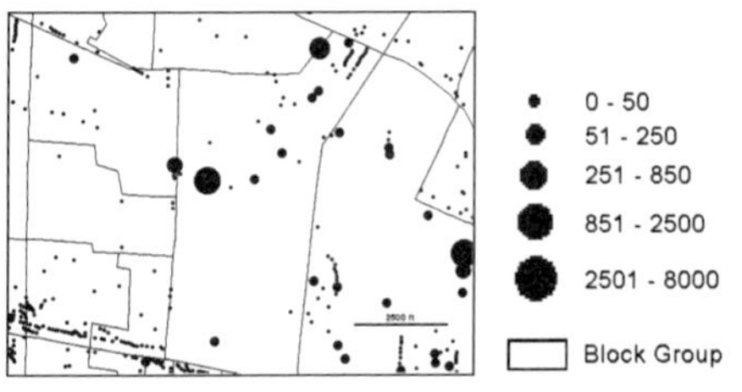

Businesses mapped by number of employees

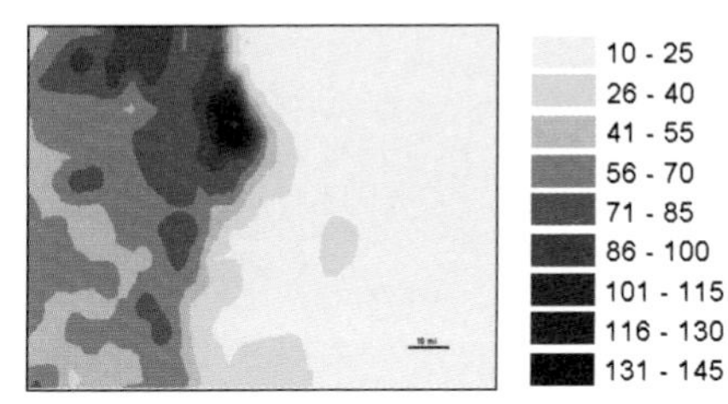

Average annual precipitation (inches)

If you're summarizing by area, using counts or amounts can skew the patterns if the areas vary in size. You should use ratios to accurately represent the distribution of features. However, you can map counts or amounts for data summarized by area if you want to focus on the quantity in each area, rather than see the patterns of where the most and least are.

For example, you might map individual businesses by number of employees, but you wouldn't map block groups by number of employees (based on locations of businesses), because block groups vary in size. Larger block groups might have more workers, but more spread out. You'd need to map workers per square mile to see the distribution.

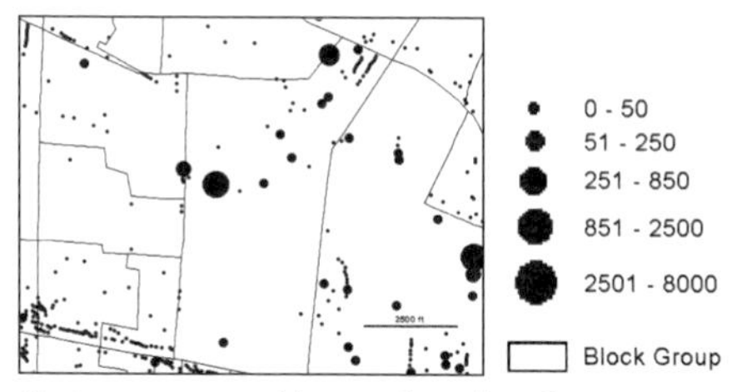

Businesses mapped by number of employees

Mapping the number of employees per block group shows the total in each.

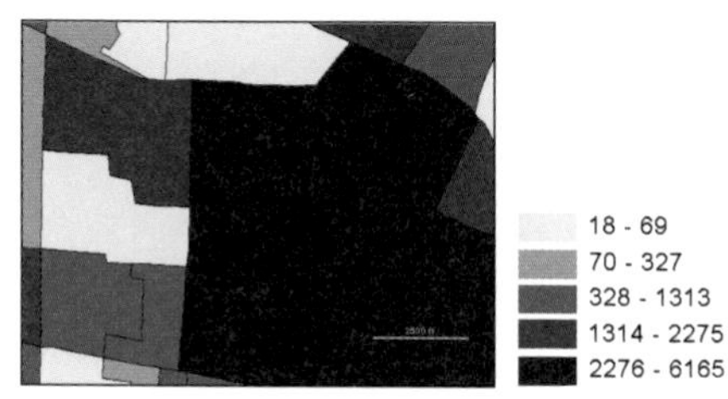

Mapping the number of employees per square mile shows the distribution.

RATIOS

Ratios show you the relationship between two quantities, and are created by dividing one quantity by another, for each feature. Using ratios evens out differences between large and small areas, or areas with many features and those with few, so the map more accurately shows the distribution of features. Because of this, ratios are particularly useful when summarizing by area.

The most common ratios are averages, proportions, and densities.

Averages are good for comparing places that have few features with those that have many. To create an average, you divide quantities that use different measures. For example, dividing the number of people in each tract by the number of households gives you the average number of people per household.

```
People per HH = Population / Households
```

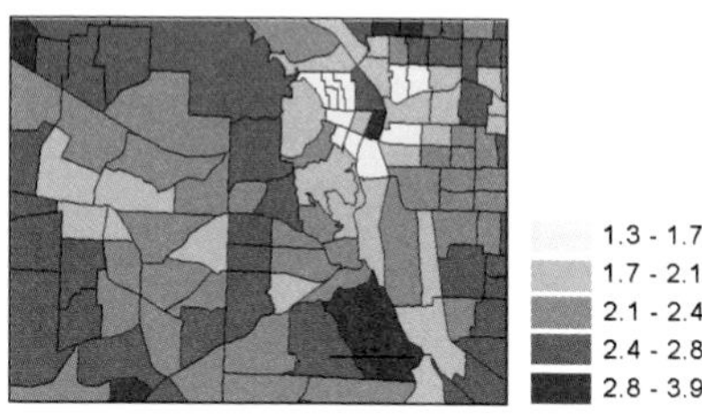

Tract	Population	Households	People per HH
003603	1606	643	2.5
0074	2765	1104	2.5
003702	2443	894	2.7
003803	4132	1591	2.6
0076	3176	1256	2.5

Number of people per household, by census tract

Proportions show you what part of a whole each quantity represents. To calculate a proportion, you divide quantities that use the same measure. For example, dividing the number of 18- to 30-year-olds in each tract by the total population of each tract gives you the proportion of people aged 18 to 30 in each tract.

`% 18-29 = 18-29 Years / Population`

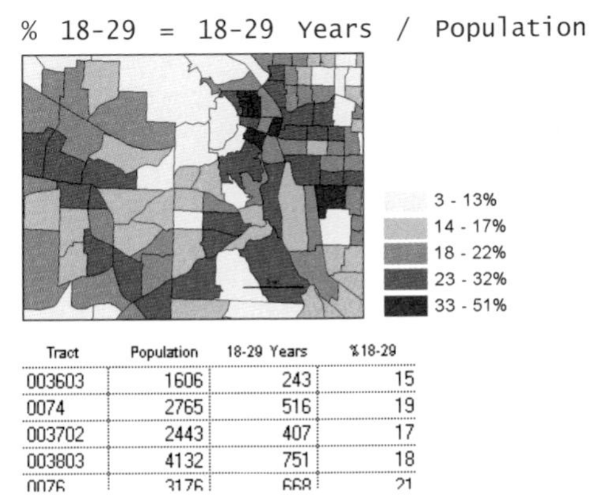

Tract	Population	18-29 Years	%18-29
003603	1606	243	15
0074	2765	516	19
003702	2443	407	17
003803	4132	751	18
0076	3176	668	21

Percentage of population aged 18 to 29, by census tract

Proportions are often presented as percentages (the proportion multiplied by 100) as a kind of shorthand. People are used to thinking and talking in terms of percentages rather than ratios—for example, "22 percent," rather than "twenty-two one-hundredths."

Tract	Population	18-29 Years	18-29 (Ratio)	18-29 (%)
031501	1853	241	0.13	13
003601	4439	799	0.18	18
003902	3072	461	0.15	15
003602	6578	1118	0.17	17

Densities show you where features are concentrated. To calculate density, you divide a value by the area of the feature to get a value per unit of area. For example, by dividing the population of a county by its land area in square miles, you get a value for people per square mile. Density is good for showing distribution when the size of the areas you're summarizing by varies greatly. For example, census tracts have roughly the same number of people, so some are small (where there are many people close together), and some are much larger (where the people are spread out). Density is the subject of chapter 4, 'Mapping density.'

`Population per Square Mile = Population / Square Miles`

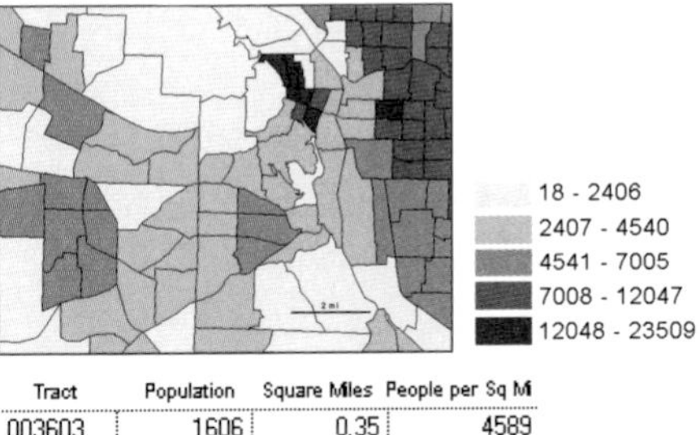

Tract	Population	Square Miles	People per Sq Mi
003603	1606	0.35	4589
0074	2765	0.58	4767
003702	2443	0.37	6603
003803	4132	0.48	8608
0076	3176	0.53	5992

Number of people per square mile, by census tract

You create ratios by adding a new field to the layer's data table, and calculating the new values by dividing the two fields containing the counts or amounts. Some GIS software, such as ArcInfo® and ArcView® GIS, lets you create ratios on the fly by doing the calculation when you create the map—you just tell it which fields you want to divide. Be careful not to create ratios from other ratios, or else you'll create values that don't really mean anything. For example, dividing the percentage of 18- to 30-year-olds in each tract by the area of the tract, to get a percentage per square mile, is meaningless.

RANKS

Ranks put features in order, from high to low. They show relative values rather than measured values. Ranks are useful when direct measures are difficult, or if the quantity represents a combination of factors. For example, it's hard to quantify the scenic value of a stream. You may be able to state, however, that the section that passes through a mountain gorge has a higher scenic value than the section passing near a dairy farm.

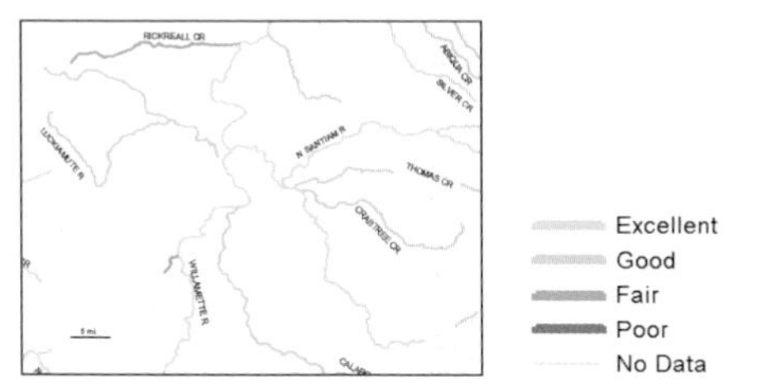

Streams ranked by recreation value

To indicate ranks, you can use text (for example, high, moderate, or low) or numbers (for example, 1 through 10). Since the ranks are relative, you only know where a feature falls in the order—you don't know how much higher or lower a value is than another value. For example, you may know a feature with a rank of "3" is higher than one ranked "2" and lower than a "4," but you don't know how much higher or lower.

You often assign ranks based on another feature attribute, usually a type or category, or a combination of attributes. For example, you'd assign all soils of a certain type the same suitability for growing a particular crop.

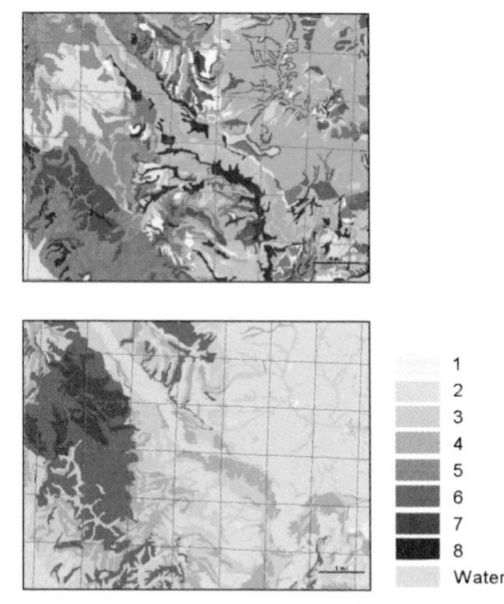

Each soil type (shown on the left) is assigned a rank based on its capability for growing crops. The highest capability is "1," while "8" is lowest.

Once you've determined what type of quantities you have, you need to decide how to represent them on the map; either by assigning each individual value its own symbol, or grouping the values into classes.

Mapping quantities involves a trade-off between presenting the data values accurately, and generalizing the values to see patterns on the map.

Usually, counts, amounts, and ratios are grouped into classes, since each feature potentially has a different value. This is especially true if the range of values is large. If each value were mapped using a unique symbol, your map would accurately reflect the data, but finding features with similar values would be difficult unless you were mapping just a few values. Using classes is especially valuable when the map will be used for public discussion, as it lets map readers compare areas quickly.

These maps show the poverty rate (percentage of people living below the poverty level) by block group. On the left, each block group is drawn using a unique gray shade based on its data value. Lighter block groups have lower values. Grouping the values into classes (right map) makes the patterns easier to see.

Ranks lend themselves to being mapped as individual values, since the values are not continuous—there is a fixed number of values, and several features are likely to have the same value.

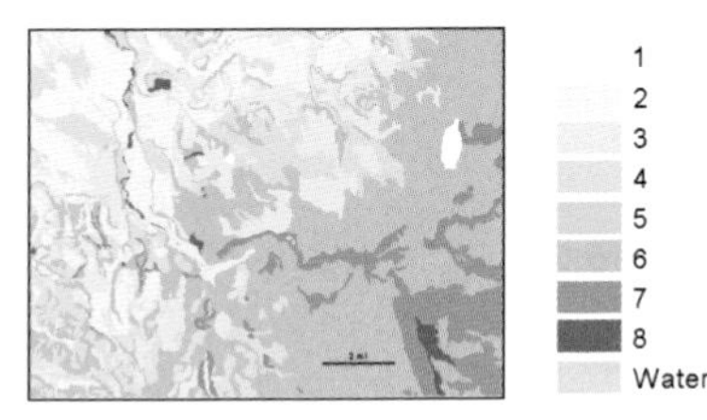

Here, soils are ranked by suitability for growing crops. Each rank is drawn with a different shade.

MAPPING INDIVIDUAL VALUES

By mapping individual values, you present an accurate picture of the data, since you don't group features together. However, this approach may require more effort on the part of map readers to understand the information being mapped, especially if the map contains many values.

Mapping individual values lets you search for patterns in the raw data. You may want to do this if you're unfamiliar with the data or area being mapped, or are looking for subtle patterns in the data. You might also do this to help decide how to group the values into classes.

If you're mapping ranks, assign one symbol to each rank. If you have more than eight or nine ranks, though, you may want to group them into classes, since too many different symbols on a map can make it difficult for map readers to distinguish the different ranks. You can do this by simply assigning the same symbol to adjacent ranks.

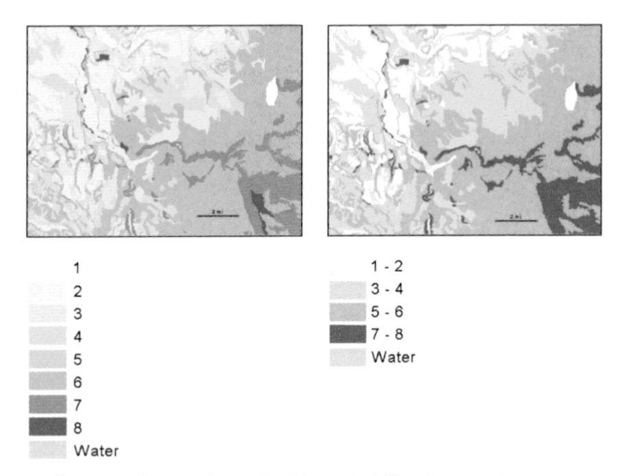

Both maps show soils ranked by suitability for growing crops. Combining the original eight ranks into four makes the patterns more distinct.

You can also map ratios, counts, or amounts using individual values if you have no more than 11 or 12 unique values, or fewer than 20 features.

USING CLASSES

Classes group features with similar values, by assigning them the same symbol. This lets you see features with similar values. How you define the class ranges will determine which features fall into each class, and thus what the map will look like. By changing the classes, you can create very different maps. Generally, make sure features with similar values are in the same class, and make the difference in values between classes as great as possible.

You can create classes manually, or use a standard classification scheme.

Creating classes manually

You should create classes manually if you're looking for features that meet specific criteria, or comparing features to a specific, meaningful value. You specify the upper and lower limit for each class and assign the symbols.

The classes might be based on a regulation that specifies a value above or below which some action occurs. If, for example, you want to find out where to designate an Urban Empowerment Zone, you'd map tracts with at least 35 percent of residents living below the poverty level. One of the class breaks for the map would be 35 percent. Another logical break would be 50 percent, to show which tracts have at least half the residents living below the poverty level.

Percentage of residents living below the poverty level, by block group

The classes might also be based on standards or research in a particular discipline or industry. For example, conservation biologists creating a wildlife corridor would exclude watersheds with less than 50 percent forest cover, and try to include watersheds with more than 85 percent forest cover. In this case, the classes would be "Less than 50%", "51 – 85%", and "Greater than 85%".

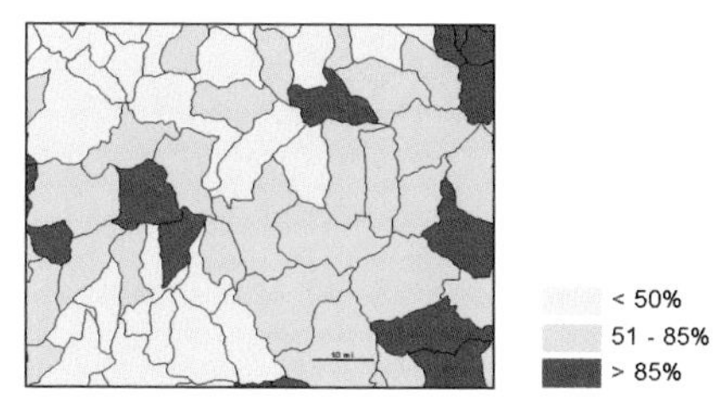

Percentage of forest cover, by watershed

You might also create classes using a value based on a larger set of features. For example, if you're mapping census tracts for a county by average number of people per household, you might make one of the class breaks the value of the national average for this statistic (2.6 in the United States). Map readers would quickly see how tracts in the county compare to the national average.

Average number of people per household, by census tract. The national average in the United States is 2.6.

In any case, you should always state explicitly on the map what the classes represent.

Using standard classification schemes

Use a standard classification scheme if you want to group similar values to look for patterns in the data. You can choose from a number of schemes for grouping data values into classes; these are based on how the data values are distributed. You specify the classification scheme and the number of classes, and the GIS calculates the upper and lower limit for each class. The four most common schemes are natural breaks, quantile, equal interval, and standard deviation.

You'll figure out the best scheme for creating the class breaks by looking at the distribution of data values. Then, you can decide on the number of classes.

A good way of seeing how data values are distributed is to plot them on a chart. In this example, the charts and maps all use the same data: median household income, in U.S. dollars, by census block group. The chart on the facing page shows the distribution of data values for the area shown on the map. Median income is plotted along the horizontal axis, and the number of block groups with each value is shown on the vertical axis. The height of the bar indicates the number of block groups with that value. The shaded areas (corresponding to the shaded classes on the map) show the range for each scheme; the values of the class breaks are indicated on the horizontal axis. The width of each area shows how many block groups fall into each class.

Natural breaks (Jenk's)

Classes are based on natural groupings of data values. On the chart, class breaks are set where there is a jump in values, indicated by a large step between bars, so block groups having similar values are placed in the same class. The resulting map emphasizes the differences between the highest income block groups, in the lower left, and the next highest, in the center.

Quantile

Each class contains an equal number of features. On the chart, the shaded areas show which block groups are in the same class, and indicate the class breaks where they cross the horizontal axis. On this map, block groups with similar values are forced into adjacent classes, and the block groups at the high end (with values ranging from $32,000 to $100,000) are lumped into one class.

Equal Interval

The difference between the high and low values is the same for every class—in this case, $20,000. On this map, almost all the block groups are contained in the two lowest classes. The map emphasizes the location of the few block groups with the very highest median income.

Standard Deviation

Features are placed in classes based on how much their values vary from the mean. The GIS calculates the mean (in this case, about $26,000) and the standard deviation (about $12,900). It successively adds or subtracts the standard deviation to or from the mean to set the class breaks. The map shows how many standard deviations each block group is from.

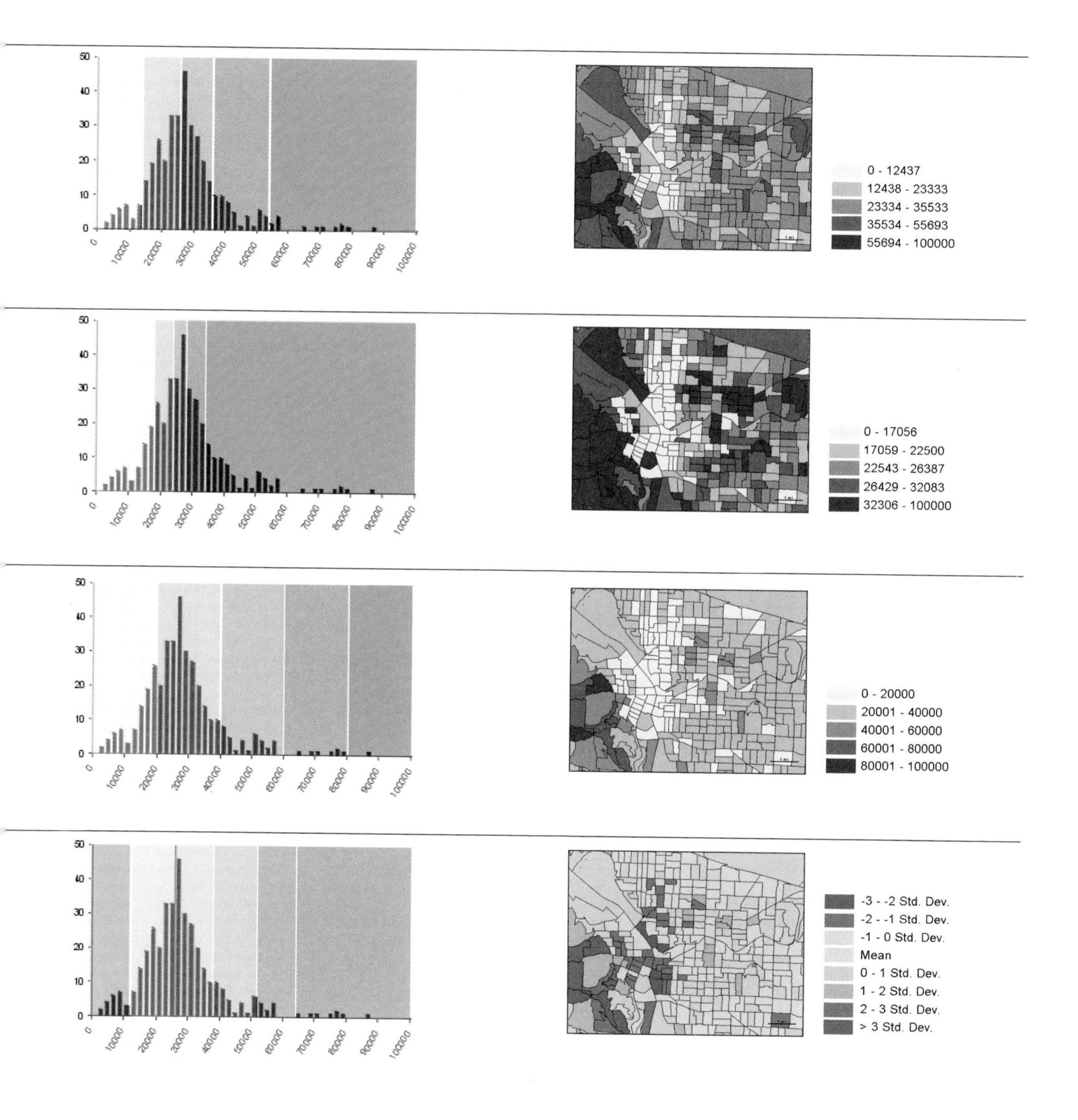

0 - 12437
12438 - 23333
23334 - 35533
35534 - 55693
55694 - 100000
0 - 17056
17059 - 22500
22543 - 26387
26429 - 32083
32306 - 100000
0 - 20000
20001 - 40000
40001 - 60000
60001 - 80000
80001 - 100000
-3 - -2 Std. Dev.
-2 - -1 Std. Dev.
-1 - 0 Std. Dev.
Mean
0 - 1 Std. Dev.
1 - 2 Std. Dev.
2 - 3 Std. Dev.
> 3 Std. Dev.

Comparing classification schemes

Natural breaks

Natural breaks finds groupings and patterns inherent in your data, so values within a class are likely to be similar, and values between classes, different. Data values that cluster are placed into a single class. Class breaks are defined where there is a gap between clusters of values.

How it works

The GIS automatically determines the high and low value for each class, using a mathematical procedure to test different class breaks. It picks the class breaks that best group similar values and maximizes the differences between classes.

What it's good for

Mapping data values that are not evenly distributed, since it places clustered values in the same class.

Disadvantages

- Since the class ranges are specific to the individual dataset, it's difficult to compare the map to other maps.
- Choosing the optimum number of classes is difficult, especially if the data is evenly distributed.

Quantile

Each class has an equal number of features in it.

How it works

The GIS orders the features, based on the attribute value—from low to high—and sums the number of features as it goes. It divides the total by the number of classes you've specified to get the number of features in each class. It then assigns the first features in the order to the lowest class until that class is filled, then moves on to the next class, fills it up, and so on.

What it's good for

- Comparing areas that are roughly the same size.
- Mapping data in which the values are evenly distributed.
- Emphasizing the relative position of a feature among other features. For example, you can show which counties in a state are in the top 20 percent, in terms of median income (those in the highest of five categories).

Disadvantages

- Features with close values may end up in different classes, especially if values cluster. This may exaggerate the differences between features. Conversely, a few widely ranging adjacent values may end up in the same class, minimizing the differences between these features.
- If the areas vary greatly in size, a quantile classification can skew the patterns on the map.

Equal interval

Each class has an equal range of values—that is, the difference between the high and low value is the same for each class.

How it works

The GIS subtracts the lowest value in the data set from the highest. It then divides that number by the number of classes you specified. It adds that number to the lowest data value to get the maximum value for the first class. It then adds to each maximum value to set the breaks for the rest of the classes.

What it's good for

- Presenting information to a nontechnical audience. Equal intervals are easier to interpret since the range for each class is equal. This is especially true if the data values are familiar to the reader, such as percentages.
- Mapping continuous data, such as precipitation and temperature.

Disadvantages

- If the data values are clustered rather than evenly distributed, there may be many features in one or two classes and some classes with no features.

Standard deviation

Each class is defined by its distance from the mean value of all the features.

How it works

The GIS first finds the mean value by adding all the data values and dividing by the number of features. It then calculates the standard deviation by subtracting the mean from each value and squaring it (to make sure it's positive), summing these numbers, and then dividing by the number of features. It then takes the square root to get the final number. The formula looks like this:

$$s = \sqrt{\frac{\Sigma(x-\bar{x})^2}{n}}$$

where s is the standard deviation, x is the value of a feature, $\bar{x}$ is the mean, and n is the number of features.

You can think of this as the average amount the data values vary from the mean. The GIS creates class breaks above and below the mean based on the number of standard deviations you specify, such as 1/2 or 1 standard deviation.

What it's good for

- Seeing which features are above or below an average value.
- Displaying data that has many values around the mean, and few further from the mean (a bell curve, or normal, distribution).

Disadvantages

- The map doesn't show the actual values of the features, only how far their value is from the mean.
- Very high or low values (outliers) can skew the mean so that most features will fall in the same class.

Choosing a classification scheme

To decide which scheme to use, you need to know how the data values are distributed across their range. Create a bar chart and set the horizontal axis to be the attribute values. The vertical axis should represent the number of features having a particular value. Most spreadsheets can create charts, as can statistical programs and GIS programs such as ArcInfo and ArcView GIS. Follow these guidelines when choosing a classification scheme.

- If your data is unevenly distributed (many features have the same or similar values, and there are gaps between groups of values), use natural breaks.

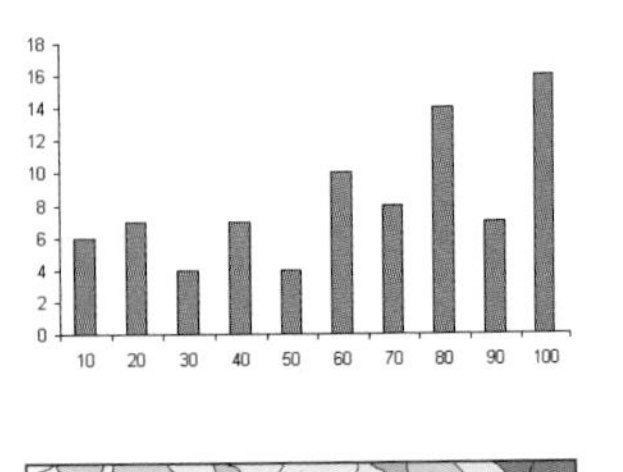

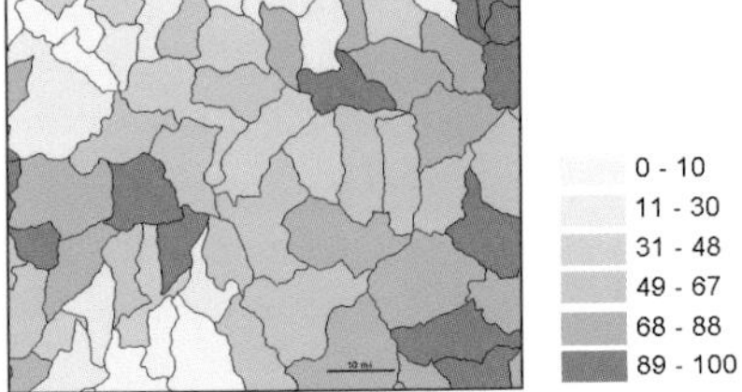

Watersheds by percentage forested. The chart shows the gaps between groups of values (the shorter bars).

- If your data is evenly distributed and you want to emphasize the difference between features, use equal interval or standard deviation.

- If your data is evenly distributed and you want to emphasize the relative difference between features, use quantile.

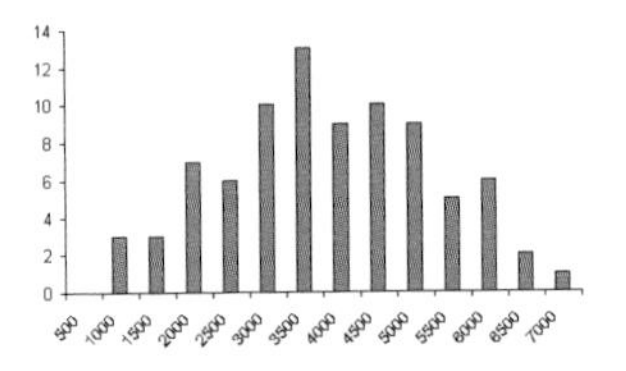

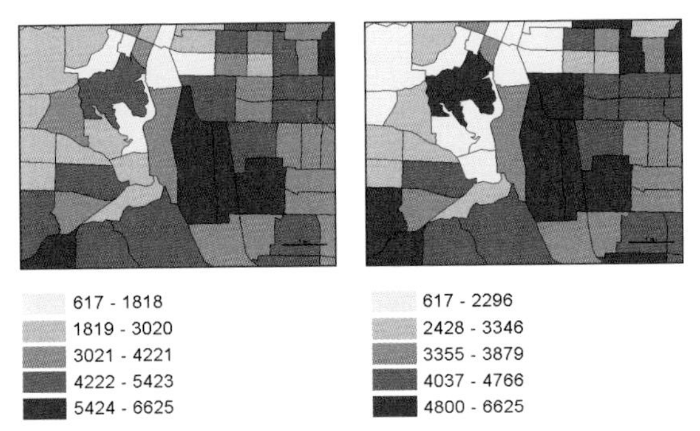

Population by census tract. The chart shows that the data values are fairly evenly distributed, with no large gaps between values. In this case, although the equal interval and quantile schemes produce similar results, equal interval places fewer features in the highest and lowest classes, emphasizing these extremes.

The GIS lets you change the class ranges, the number of classes, and the symbols you use to represent the features fairly quickly, so you can try several approaches to see which one communicates the information best. This is especially useful if you're exploring the data and searching for patterns.

Dealing with outliers

When you graph the data, you may find that you have a few extremely high or low values. These outliers can skew your class ranges, and hence the patterns on the map. This is particularly true when using the equal interval or standard deviation schemes, since all features, except the outlier, may be forced into a single class. Using natural breaks can isolate outliers in the highest or lowest class, but can still compress values into the remaining classes.

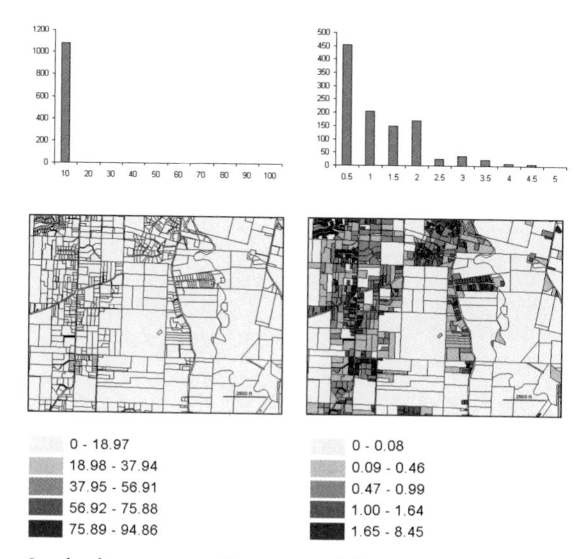

Land value per square foot, in U.S. dollars. In this example, an error in the database resulted in one extremely high data value, $94.86 per square foot (left chart and map). Using an equal interval classification, all features are in the lowest class, the outlier is in the highest, and the remaining classes have no features. With the error corrected, the bar chart and map show a more even distribution of values.

You should look at outliers closely. They may be the result of an error in the database, or they may be anomalies based on a small data sample. Or, they may be completely valid. If they're not outright errors that can be corrected in the database, you can handle them in several different ways, depending on how much they vary from the rest of the values, and how they affect the patterns on the map:

- Put each outlier in its own class. You might do this if the outliers are widely spread out.
- Group them together into a class. You might do this if the outliers cluster.
- Group them with the next closest class up or down, if they're not too far from other values in the class.
- Draw them using a special symbol, if you believe they're not really valid and should not be considered part of the pattern. For example, you might shade them in gray and label them as "Insufficient Data" in the legend.

Outliers can occur as the result of a ratio calculation. While each of the original values might be valid on their own, when divided, the resulting value might be misleading. This could be caused by the way features are stored in the database, or the relationship between the values you're mapping. Suppose, for example, you're mapping the number of grocery stores in each census tract, per 1,000 people. If there's a store in a tract where few people live (perhaps an area with many office buildings, but few houses), calculating stores per 1,000 people for this tract will result in a very high value, skewing the classes and obscuring the patterns in the map.

```
Stores per 1000 people =
# of Stores / (Population / 1000)
```

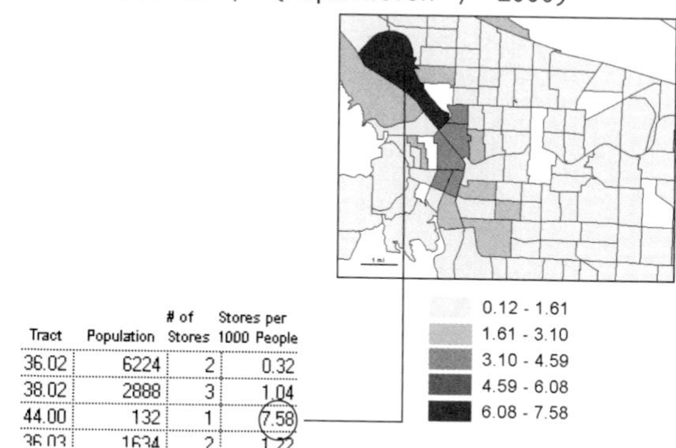

Tract	Population	# of Stores	Stores per 1000 People
36.02	6224	2	0.32
38.02	2888	3	1.04
44.00	132	1	7.58
36.03	1634	2	1.22

Number of grocery stores per 1,000 people, by census tract. The small population in tract 44.00 creates an outlier when the number of stores per 1,000 people is calculated.

Deciding how many classes

Once you've decided on an appropriate classification scheme, you need to decide how many classes to create. Based on this number, and the classification scheme, the GIS calculates the class ranges and breaks. If you've chosen an appropriate classification scheme, changing the number of classes shouldn't change the appearance of the data very much—only make the patterns more or less distinct.

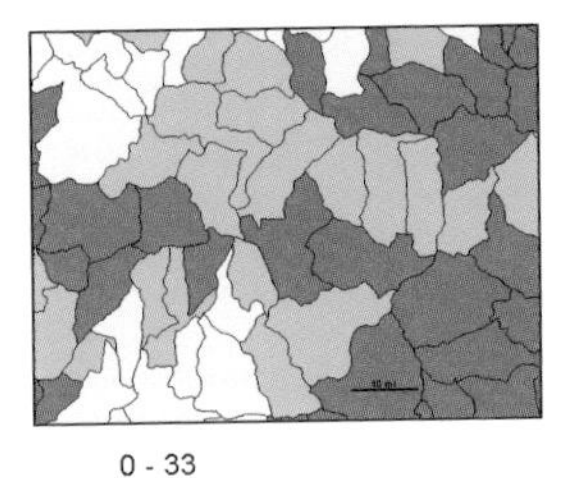

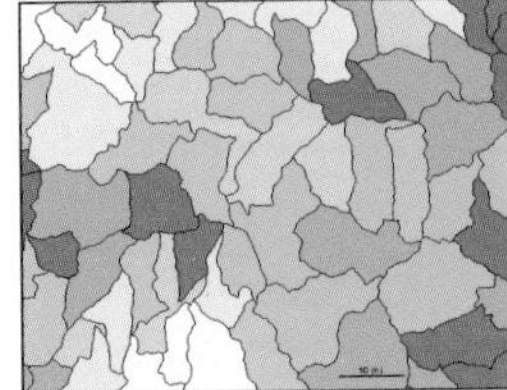

Watersheds by percentage of forest cover. Using four classes makes the patterns distinct; using more classes reveals subtleties in the patterns.

- Most map readers can distinguish up to seven colors on a map, so using more than seven classes will make it hard to find features with similar values. Four or five classes will usually reveal patterns in the data, without confusing the reader. Using fewer than three or four classes doesn't show much variation between features, and hence shows no clear patterns.

- If you're exploring the data to see what kinds of feature groupings and patterns emerge, you may want to use more classes at first. Each feature will be in a narrower range, with values closer to its actual value.

Making the classes easier to read

Once the GIS has defined the ranges, you may want to adjust them to make the classes easier to interpret quickly.

If you don't have to show the exact data values, rounding the minimum and maximum values for each class can make the legend easier to read, without changing the patterns in the map.

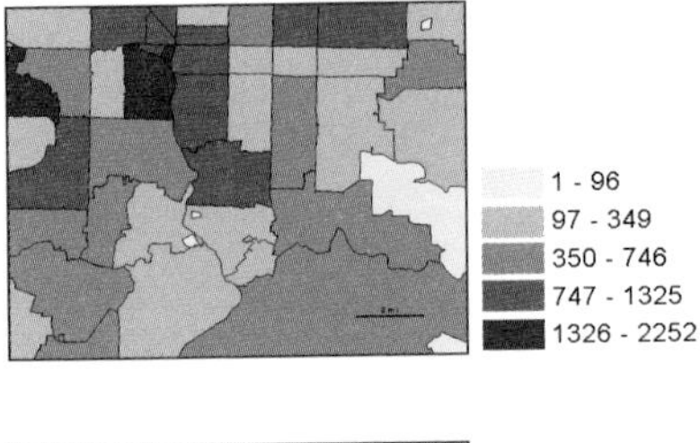

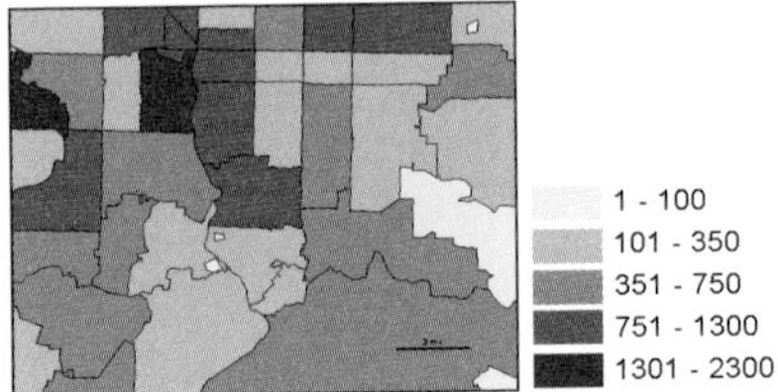

Number of businesses per ZIP Code. Rounding the values makes the numbers easier to read.

Some GIS software creates continuous class ranges by default, making the maximum value in each class the minimum of the next higher class. In fact, the lowest actual data value for the next higher class may be well above the low value shown in the legend for that class. Have the GIS define the classes. Then, change the lowest value for each class to match the lowest value of its features. While the patterns on the map won't change, the legend will better reflect the actual value ranges. This is especially true if you're using a natural breaks classification. However, you should avoid doing this if you're using an equal interval classification, since these ranges are continuous by definition.

You can rename classes in the legend to emulate ranked values, such as "very high," "high," "moderate," "low," or "none." This can make your map easier to understand quickly. You may want to do this if the relative values are more important than the actual values. This is often the case with ratios or large numbers. For example, when calculating the number of grocery stores per 1,000 people for each census tract, you may end up with decimal values that are meaningless on their own—they are only important relative to the other values. To make the map easier for readers to understand, you would change the numeric values to labels such as "high," "medium," and "low."

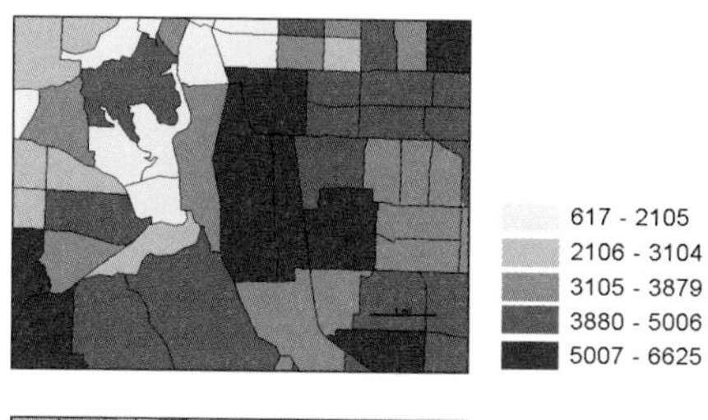

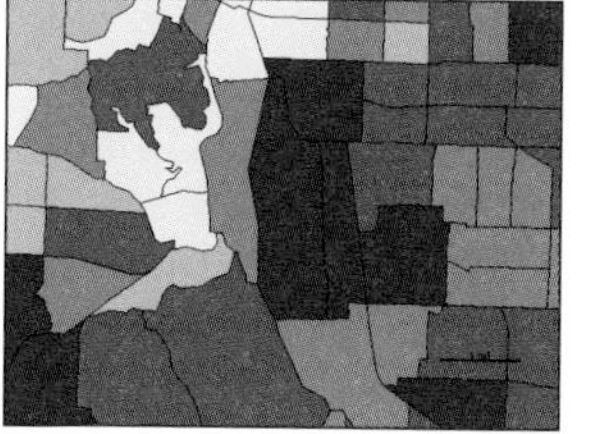

Population per census tract. Using noncontinuous ranges presents a more detailed picture of the data.

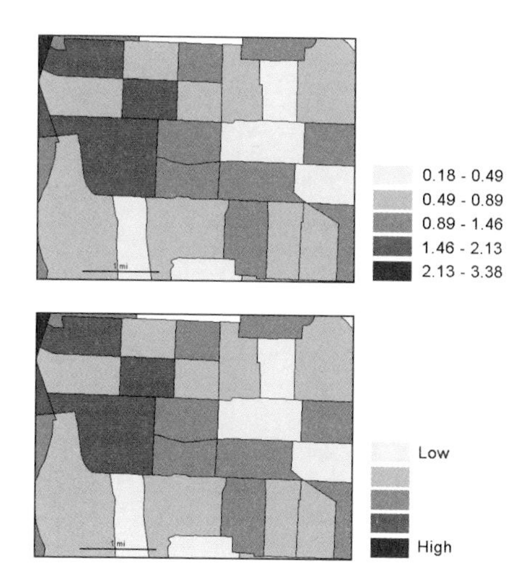

Once you've decided how to classify the data values, you'll want to create a map that presents the information to map readers as clearly as possible. Keep the map simple and present only the information necessary to show patterns in the data. Because GIS makes it easy to create maps and the database often has so much available information, there is a temptation to present more information on the map than the reader can readily comprehend.

The GIS gives you a number of options for creating maps to show quantities:

- Graduated symbols
- Graduated colors
- Charts
- Contours
- 3-D perspective views

Graduated symbols

Graduated colors

Charts

Contours

3-D perspective views

Which Features	Which Values	Advantages	Disadvantages
Locations Lines Areas	Counts/Amounts Ratios Ranks	Intuitive—people associate symbol size with magnitude	May be difficult to read if many features on map
Areas Continuous phenomena	Counts/Amounts Ratios Ranks	Makes it easy to read patterns and feature values	Colors not intuitively associated with magnitude
Locations Areas	Counts/Amounts Ratios	Shows categories as well as quantities	May present too much information, obscuring patterns
Continuous phenomena	Amounts Ratios	Easy to see rate of change across an area	May make it hard to read patterns and individual feature values
Continuous phenomena Locations Areas	Counts/Amounts Ratios	High visual impact	May make it hard to read values of individual features

CHOOSING A MAP TYPE

The option you choose depends on the type of features and data values you're mapping.

If you have discrete locations or lines, use

- Graduated symbols to show value ranges
- Charts to show both categories and quantities
- A 3-D view to show relative magnitude

If you have discrete areas, or data summarized by area, use

- Graduated colors to show value ranges
- Charts to show both categories and quantities
- A 3-D view to show relative magnitude

If you have spatially continuous phenomena, use

- Graduated colors to show value ranges
- Contours to show the rate of change
- A 3-D view to show relative magnitude

USING GRADUATED SYMBOLS

Use graduated symbols to map discrete locations or lines. Graduated point symbols are drawn at the locations of individual features to show the magnitude of the data value. Graduated line symbols are used to show the volumes or ranks for linear networks, such as roads, utility lines, or rivers.

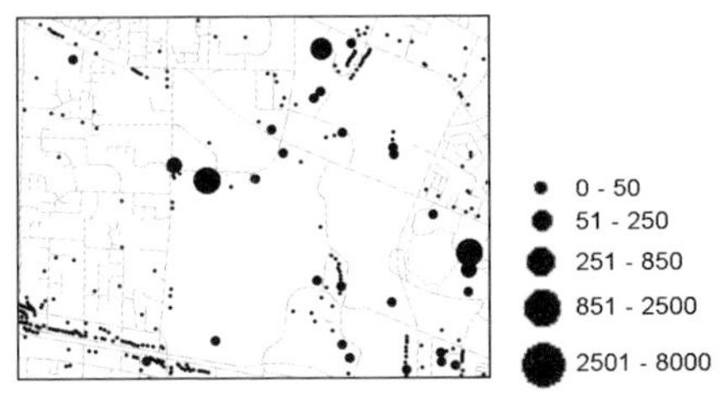

Locations—number of employees per business

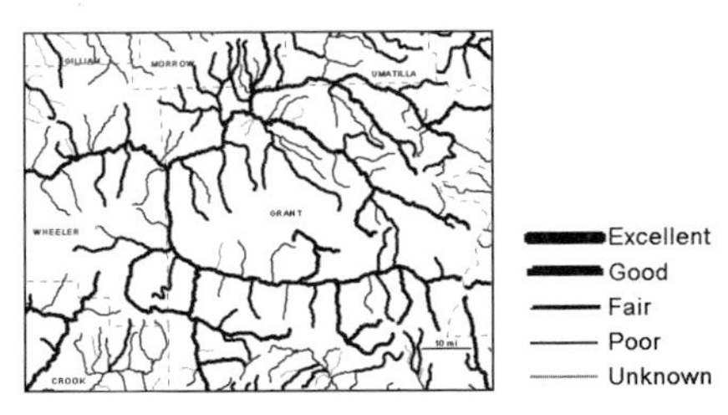

Lines—rivers ranked by fish habitat

If you're using graduated symbols with classes, you specify the minimum and maximum sizes for the symbol, and the number of classes; the GIS figures out the size of the intermediate symbols.

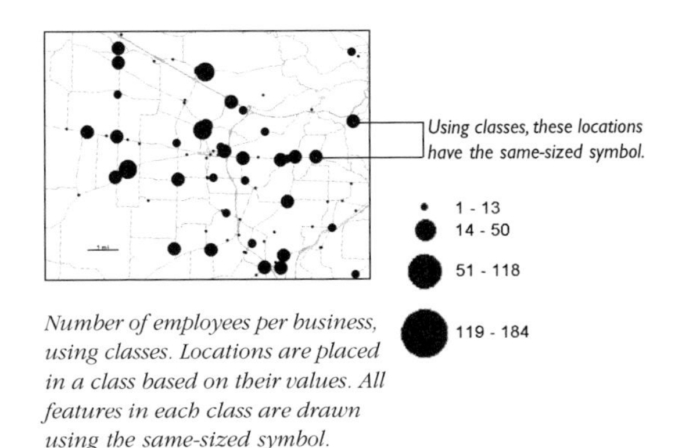

Number of employees per business, using classes. Locations are placed in a class based on their values. All features in each class are drawn using the same-sized symbol.

If you're using graduated symbols with individual values, you specify one data value and a corresponding symbol size, and the GIS scales all other symbols accordingly. The legend displays a subset of values and symbols to indicate the relative value of individual features. For example, when mapping the number of employees at each business, the legend displays four circles representing businesses with 1, 10, 50, and 100 employees. By comparing these with the circles on the map, you can tell approximately how many employees a particular business has.

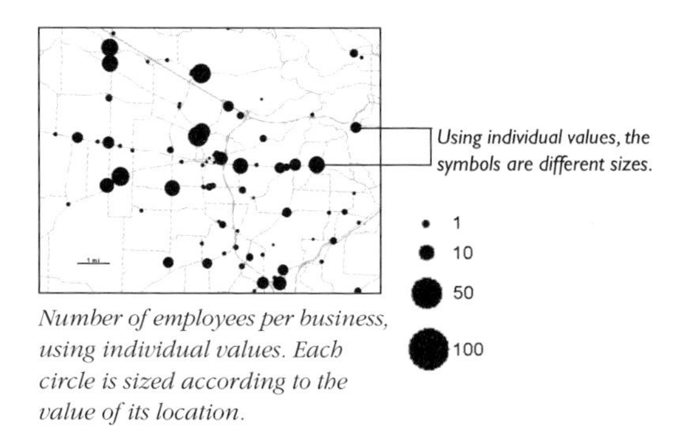

Number of employees per business, using individual values. Each circle is sized according to the value of its location.

- Most commonly, circles are used for individual locations, although any symbol can be specified. Most readers can interpret relative magnitude more easily from circles than from any other symbol.
- All the symbols should be the same color. Make sure they're dark enough to be seen easily. If you're drawing them inside areas, use a light color to shade the areas.
- Make the difference between the largest and smallest symbols great enough to show the difference in data values.

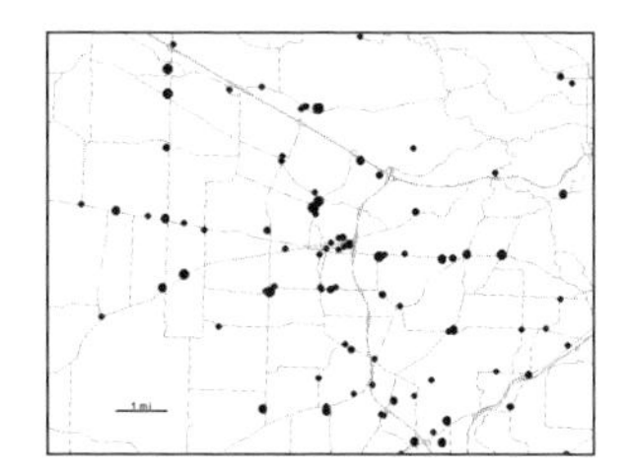

Symbols are too close to the same size to show patterns.

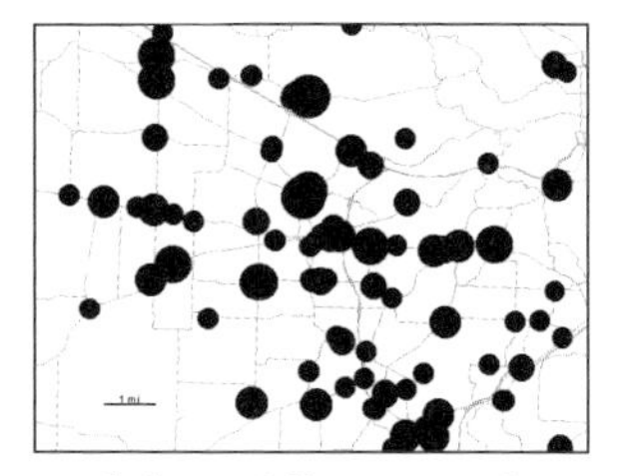

Symbols are different enough to show patterns, but obscure individual locations.

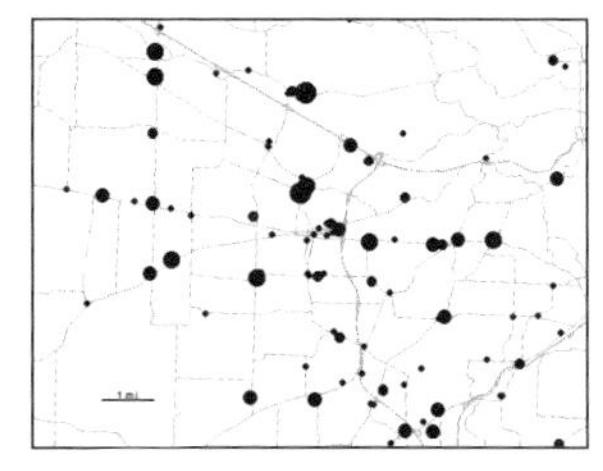

Symbols show patterns without obscuring feature locations.

- Make sure that the smallest symbols will be visible at the scale you're using, and that the largest symbols don't overlap so much that they obscure patterns on the map.

USING GRADUATED COLORS

Use graduated colors to map discrete areas, data summarized by area, or continuous phenomena. Usually, you assign shades of one or two colors to the classes. You pick the colors representing the lowest and highest classes, then pick the intermediate colors or let the GIS pick them.

If you have fewer than five or six classes, use one color and vary the shade. Remember that most people can only distinguish up to seven colors. Most people also interpret darker colors to mean "more" or "greater," so assign the darkest shade to the highest class.

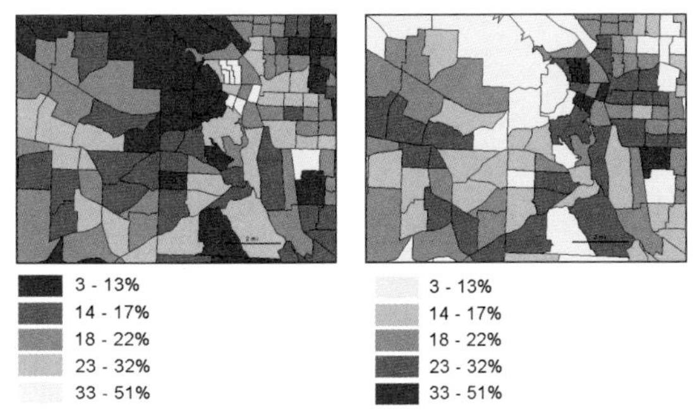

Percentage of population aged 18 to 29, by census tract. People generally associate darker colors with higher values, so the map on the left may initially be misleading.

Different colors have different visual impacts. Reds and oranges attract the most attention; blue-green, the least.

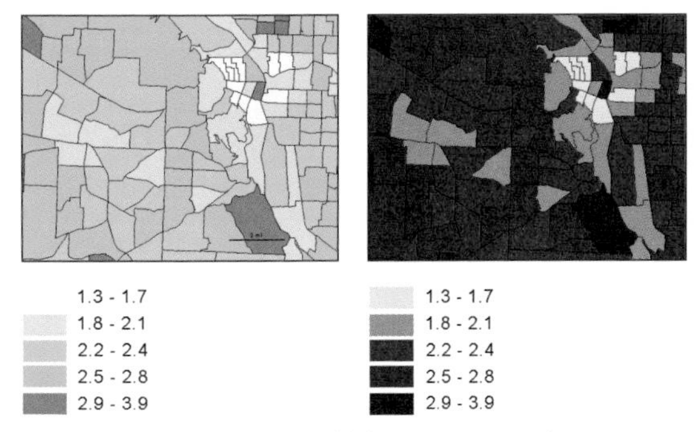

Number of people per household, by census tract. The contrasts between the reds are greater, highlighting the high and low values.

It's easier to distinguish between shades of purples and blues than shades of other colors, so you might use these if you have more than four or five classes. Keep in mind that certain colors have special meanings for some people. For example, red is often used to indicate hot spots, such as areas with many crimes, or areas unsuitable for a particular use, such as those too steep to build on.

If you have more than seven or eight classes, you may want to use a combination of colors and shades, using two or even three colors (blue to orange, or blue to green to yellow) to help distinguish the classes. Warm colors (red, orange, or yellow) are a good choice for the classes representing higher values since they highlight these values; cool colors (green, blue, or purple) can be used for lower values.

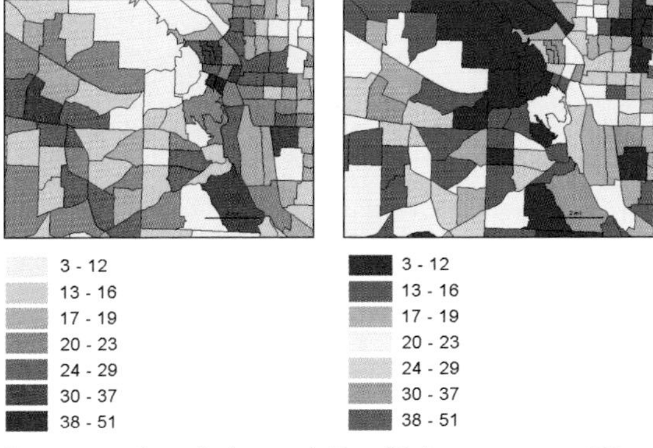

Percentage of population aged 18 to 29, by census tract. Using shades of two or more hues helps distinguish the classes.

Using two colors is also good for showing data with both positive and negative values, such as percentage above or below an average value: shades of one color (such as red) would show percentages above the average, while shades of another color (such as blue) would show percentages below, with a neutral color showing the average. This approach is particularly good for maps using classes based on standard deviation.

A narrow range of values spanning several adjacent classes may end up represented as quite different shades on the map, especially if your data is not evenly distributed. This is one reason it's important to try to make classes, and their symbols, reflect the actual distribution of data values.

USING CHARTS

Use charts to map data summarized by area, or discrete locations or areas. With charts, you can show patterns of quantities and categories at the same time (see chapter 2, 'Mapping where things are'). That lets you show more information on one map, rather than showing each category on its own map. For example, if you're mapping population by county, you can use a pie chart to show the percentage of the population by ethnic group for each county.

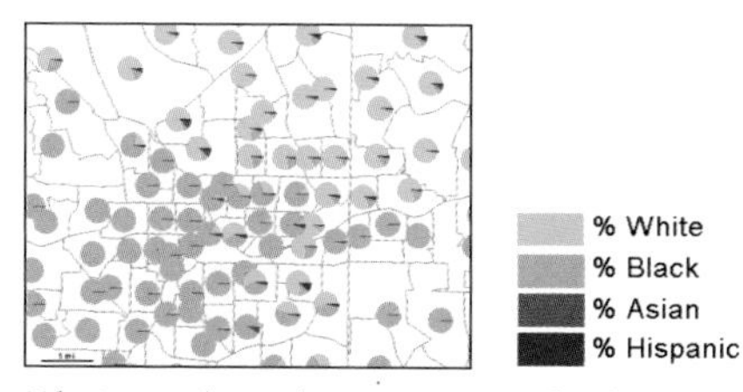

Ethnic population by census tract. The charts show a clear trend from upper right to lower left.

Percentage White

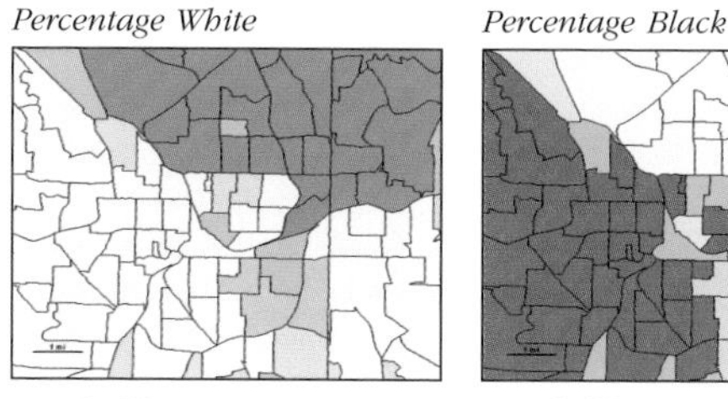

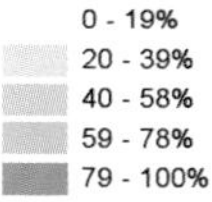

Percentage Black

0 - 20%
21 - 40%
41 - 60%
61 - 80%
81 - 100%

Percentage Asian

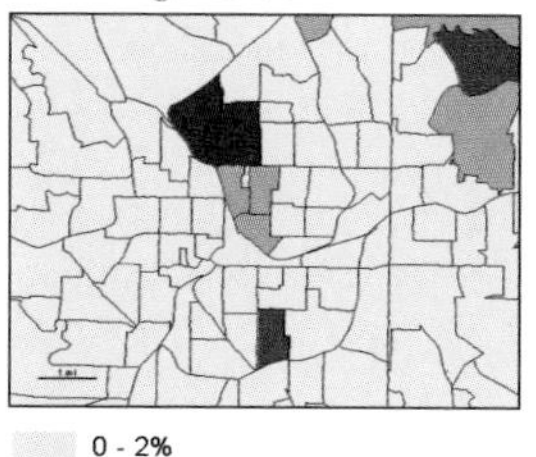

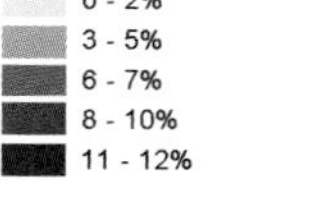

Percentage Hispanic

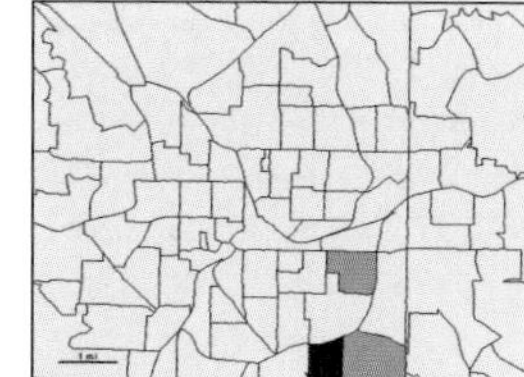

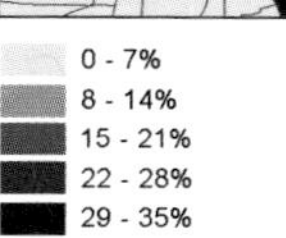

These four maps using graduated colors show more detailed patterns than the map using pie charts.

Charts are useful for a quick study of the patterns. However, some patterns may not be as readily apparent as when presented on separate maps. Maps with charts also require more effort to interpret, since you're looking at both quantities and categories together.

You can create pie charts or bar charts.

- Use pie charts if you want to show how much of the total amount each category takes up. You specify the categories and the attribute to use as the "total" value. The GIS calculates the percentage for each category and shades the chart accordingly.

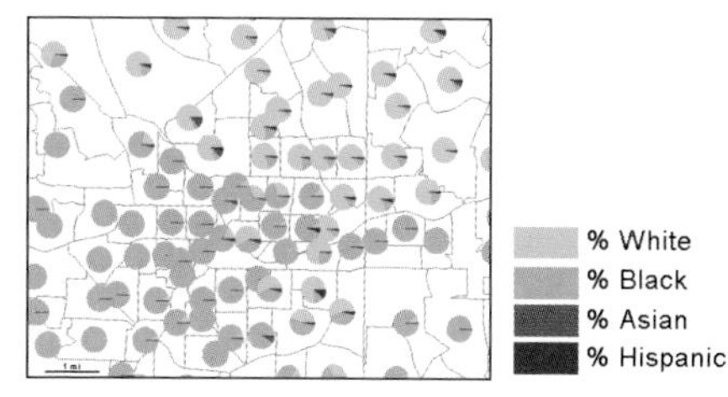

Pie charts are good for showing percentages.

- Use bar charts to show relative amounts, rather than a proportion of a total. You specify a minimum and maximum bar height, and each category is drawn according to its value.

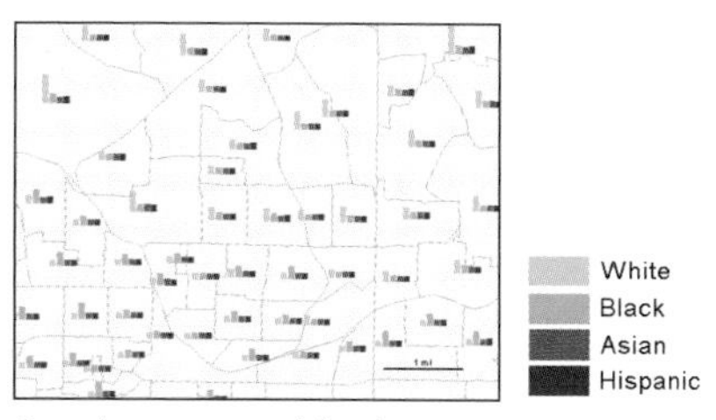

Bar charts are good for showing amounts.

You can make pie charts the same size, or vary the size according to the total amount of the attribute. For example, you could draw the charts larger or smaller, based on the total population of each census tract.

Keep the charts the same size if you want to focus on the amount of each category relative to the total.

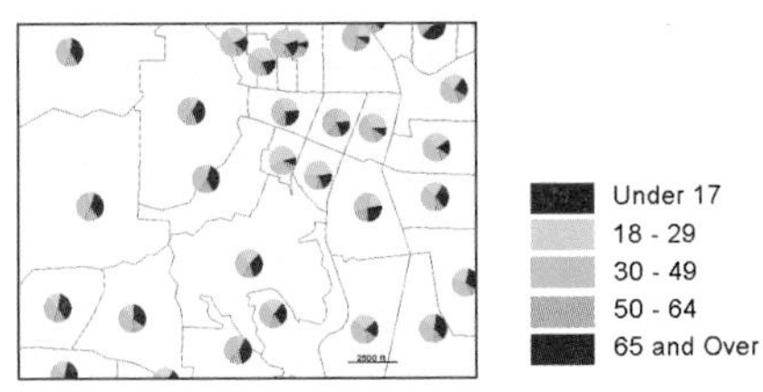

Population of census tracts by age category

Use graduated charts to show the relative size of each feature.

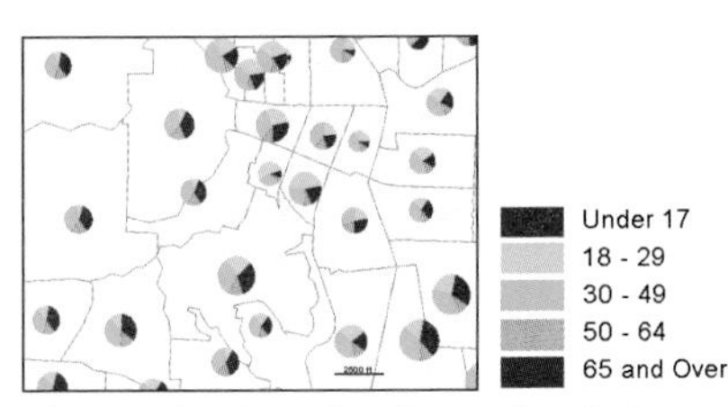

The size of each circle indicates the relative population of each tract.

Since the charts represent categories, not relative amounts, draw the bars or wedges using different colors, rather than shades of one color.

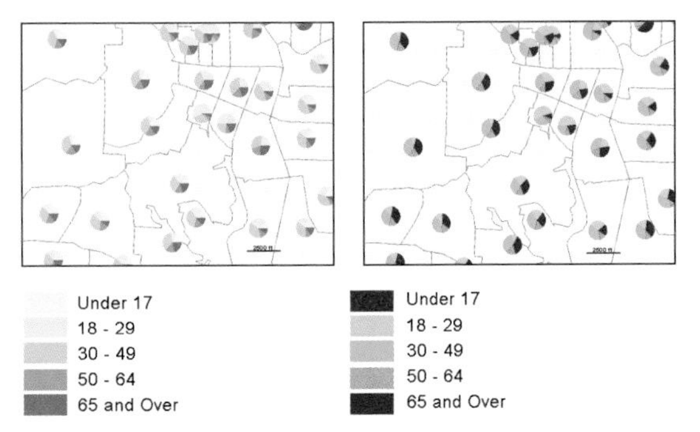

It's easier to distinguish the categories when using different colors, rather than shades of one color.

Charts are most effective when mapping no more than 30 features. Otherwise, the patterns on the map will be difficult to see.

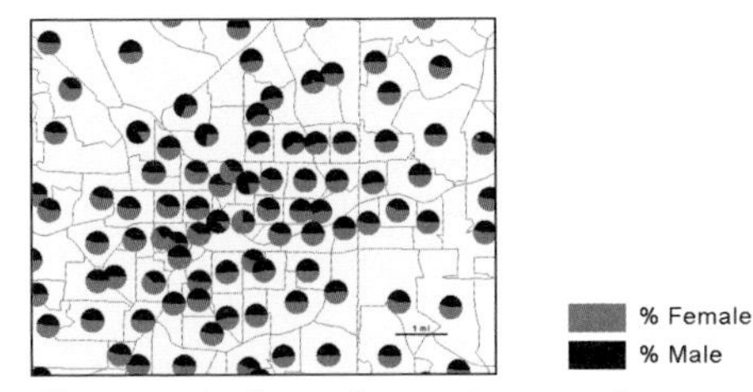

Charts aren't effective for mapping many features; the patterns are hard to see.

Don't use more than five categories on a chart; if you want to show more categories, a series of shaded maps showing each category will work better.

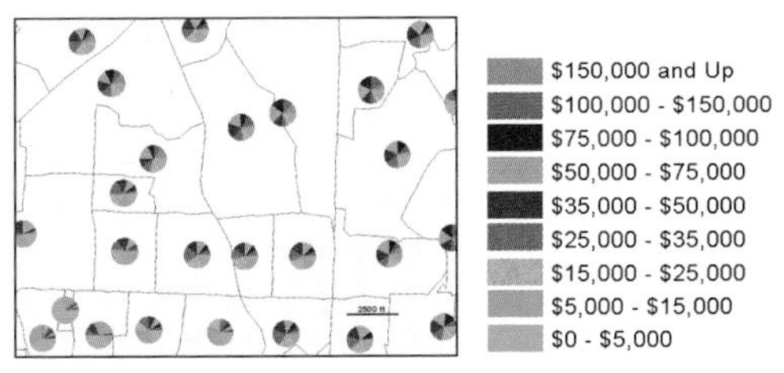

Median income, by census tract. Using too many categories makes the map hard to interpret.

Make sure the charts are large enough to read at the scale you're using. If you're using graduated charts, make sure the smallest one is big enough to be readable, and the biggest ones don't obscure the area boundaries or overlap each other.

USING CONTOUR LINES

Use contour lines to show the rate of change in values across an area for spatially continuous phenomena. Where the lines are closer together, the change is more rapid. Elevation and barometric pressure are commonly mapped using contours.

Contour lines are drawn at an interval that you specify. For instance, a contour map of precipitation with a contour interval of 10 inches would have contour lines at 10, 20, 30, and so on. Each point on a line has the same value, while a point between the lines has a value between the values of the two lines on either side of it.

The interval determines the number of lines and the distance between them. When you use a smaller interval, you create a map with more lines.

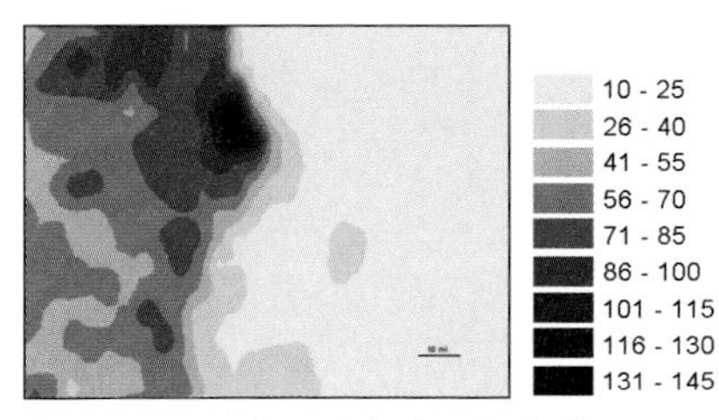

Average annual precipitation, in inches, as a surface

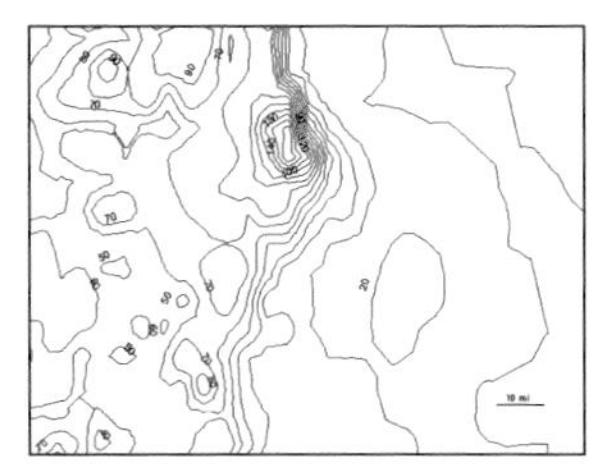

Average annual precipitation, in inches, as contour lines

- Choose an interval for the contour lines that is small enough to give some definition to the surface but not so small that the lines become too close together and the map difficult to read.

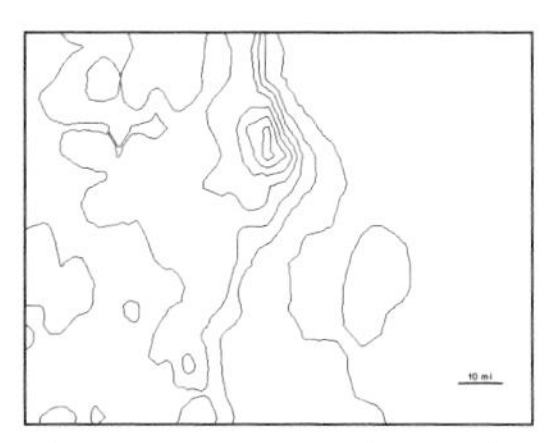

The contour interval is too large to show definition in the surface.

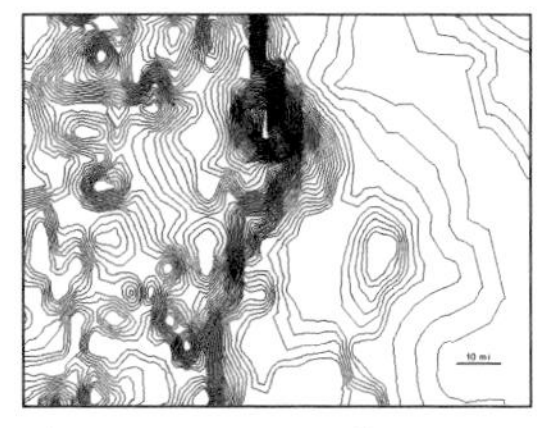

The contour interval is too small to read lines clearly.

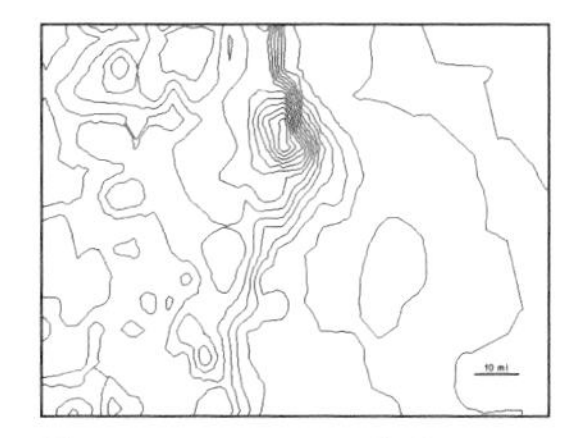

The contour interval shows definition in the surface, while making individual lines easy to see.

- Contour lines should be labeled with their value to make it easier to see the actual values, as well as the patterns.

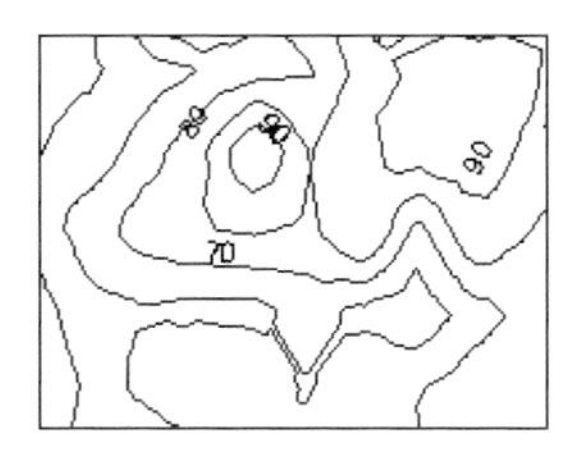

- Using a bold line for every fifth interval makes the values easier to read. For example, if the interval is 10 inches, you'd use bold lines for values of 50, 100, and 150 inches.

CREATING 3-D PERSPECTIVE VIEWS

Three-dimensional (3-D) perspective views are most often used with continuous phenomena to help people visualize the surface. You can also create 3-D views for areas or points—the height of the feature indicates the magnitude of the location or area.

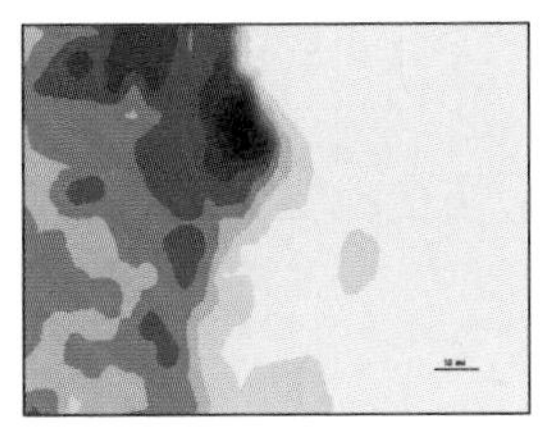

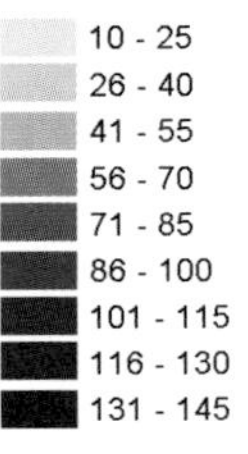

Average annual precipitation, as a surface (left), and as a 3-D perspective view (right)

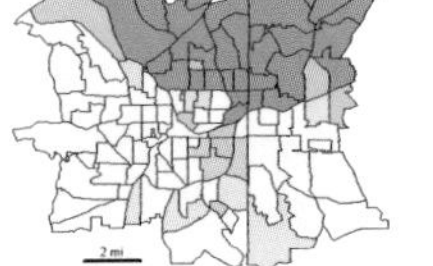

0 - 19%
20 - 39%
40 - 58%
59 - 78%
79 - 100%

Percentage white population by census tract, using graduated colors (left) and displayed as a perspective view

Although you can see where values are higher or lower when using 3-D views, it's difficult to determine the value at any particular location.

To create a 3-D view you specify three parameters that determine what your view will look like: the viewer's location, the vertical exaggeration (z-factor), and the location of the light source.

Viewer location

The viewer's location determines which features are visible in the view, since taller features may block features behind them. You set the location by rotating the view until you have the viewing position you want. Or, you specify the coordinates of the viewer and a target location on the view (usually the center of the features you're displaying), along with the viewer's angle above the surface.

Since you might rotate the view off a north–south axis to highlight patterns in the data, it's important to orient the viewer by providing a north arrow, or drawing recognizable features on the view (such as boundaries or major roads).

Rotated 180 degrees (right view), the 3-D perspective of census tracts doesn't show the steep drop-off in white population.

z-factor

When creating the view, you can specify a value, called a "z-factor," to increase the variation in the surface so that the differences are easier to see. This value is multiplied by each feature's data value. For example, using a factor of 2, a tract that has 40 percent white population will have a value of 80, and a tract that is 10 percent white will have a value of 20. The original difference between the values (30) is now 60, so the difference in heights in the view will also be greater. The goal is to use a z-factor large enough to show variation in the surface but not overly exaggerate the differences between the values.

If the vertical exaggeration is too small (right), the view doesn't show enough variation.

Light source

The location of the light source, when combined with the z-factor, determines how shadows will appear on the surface, and thus how distinct the features on the surface will be. You specify two values for the light source: the direction and the angle.

The direction the light is coming from is usually specified in degrees (from 0 to 359, with 0 being north). Unless the location of the shadows is important (for example, you're mapping terrain and want to see which areas are in shadow at a particular time), you can set the direction to highlight the features in the view. This may take some experimenting.

The angle is the height of the light source above the horizon, specified in degrees. The lower the angle, the longer the shadows.

Light source from the northwest

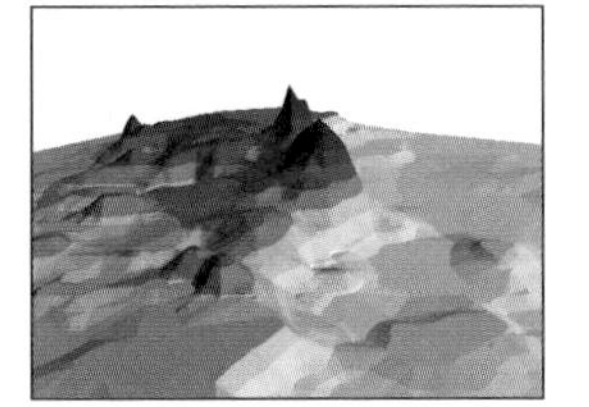

Light source from the southeast

Displaying a perspective view

If you're mapping continuous data, you'll typically draw a perspective view using a range of colors or shades. You can also draw contour lines over the view, and label them to give more definition to the surface and show data values.

You can display individual locations or areas using a single color, or you can use colors based on categories. The bottom view shows not only where the employment centers are (the clusters of taller pillars), but also which types of businesses have the largest numbers of employees.

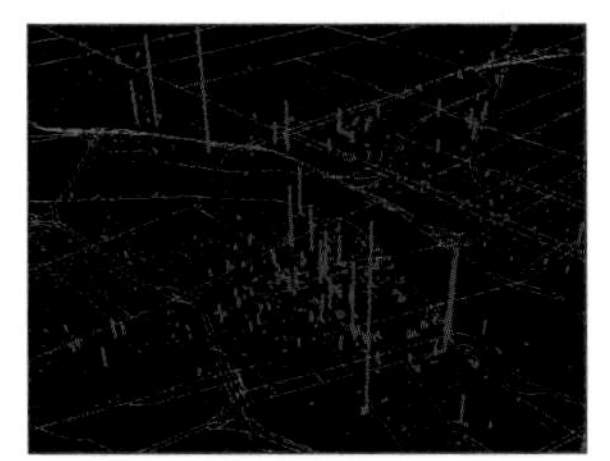

The height of the pillars represents the number of employees at each business.

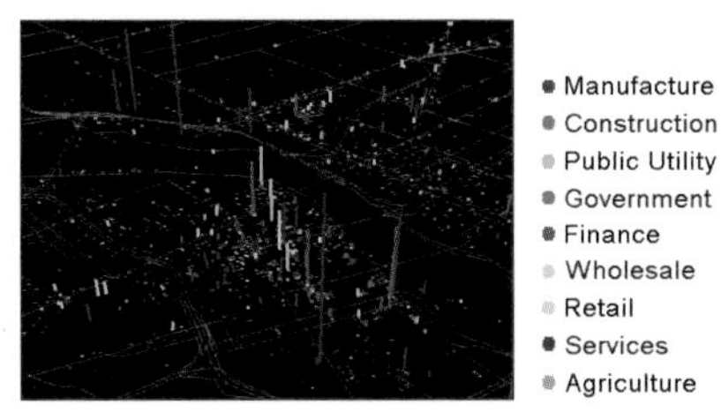

Color coding the pillars by type of business shows which types have the most workers.

If your map presents the information clearly, you can compare different parts of the map to see where the highest and lowest values are. Looking at the transition between where the least and most are—for example, seeing where change is rapid or gradual—can give you further insight into relationships between places.

You'll want to see whether values cluster or are evenly distributed. In this map, the Asian-American population is clustered in three areas. A store owner selling to this population would focus on these areas for an ad campaign.

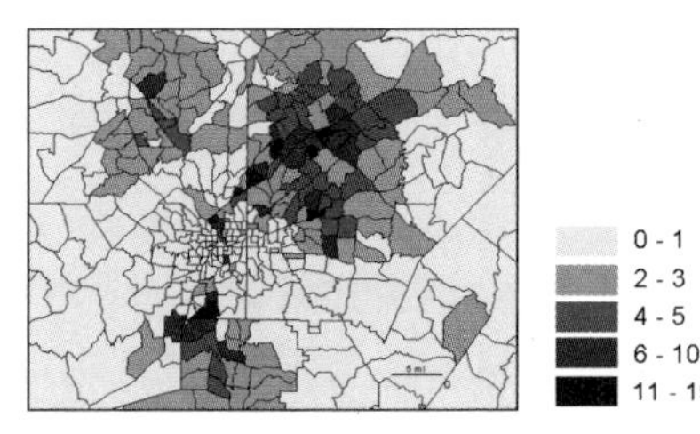

Percentage Asian-American, by census tract

Values may be concentrated in one place, or scattered across an area. In the top map (below), the African-American population is concentrated in five census tracts, and gradually diminishes outward from these. In contrast, the Hispanic population in the bottom map is fairly evenly distributed. A business targeting these customers could site a new store accordingly.

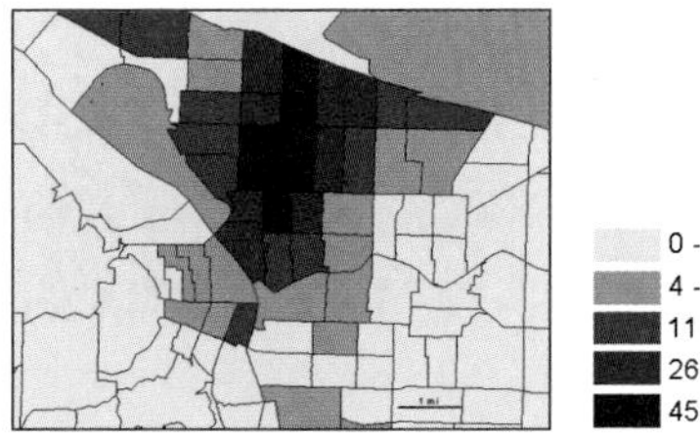
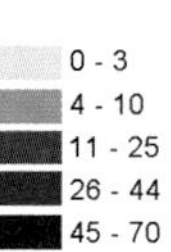

Percentage African-American, by census tract

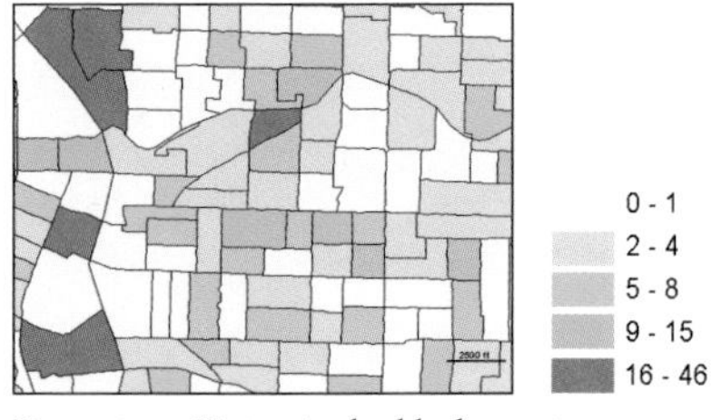

Percentage Hispanic, by block group

The relationships between the locations of the features with high and low values can help you understand how people or phenomena behave. There may be a gradual trend across the area, from low to high, as shown in the top map (percentage Native American), or there may be a definite line dividing areas of high and low values, as appears in the bottom map (percentage White). Sociologists might gain insight from these maps about the different ways ethnic groups mix in these places.

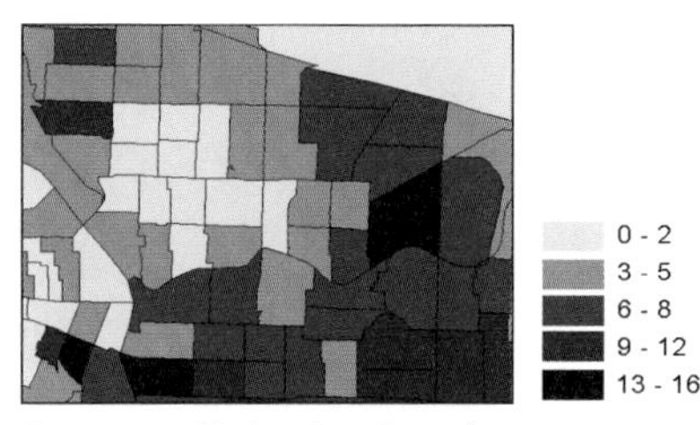

Percentage Native American, by census tract

Percentage White, by census tract

The change in values across the area may be abrupt in some areas and gradual in others. This surface map of household income might help a marketing firm delineate areas for targeted advertising.

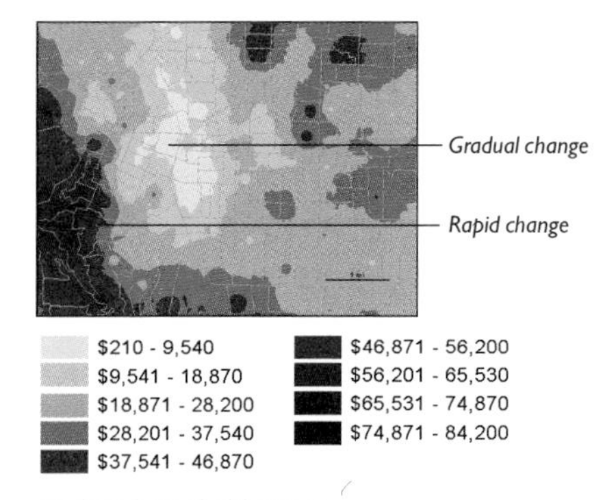

Median household income

Some features may stand apart from surrounding features. On the map below, most businesses have few employees, with one exception. A transportation planner would want to know where exceptions to the pattern of workers are.

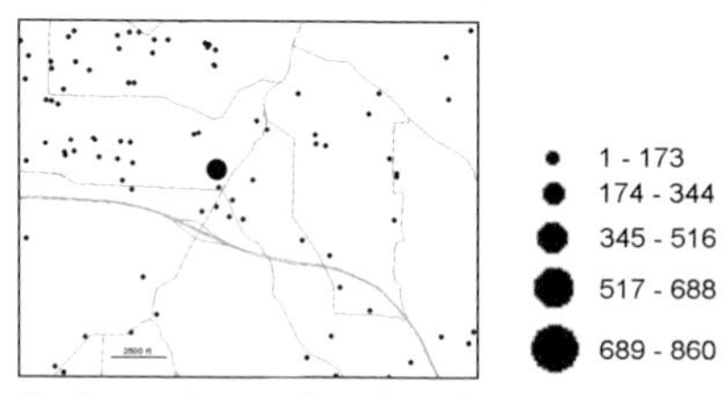

Businesses by number of employees

If you're summarizing by area, the areas you use can affect the patterns you see on the map. Using a few large areas may obscure subtle patterns. Conversely, using many small areas may present too much local variation to see the overall patterns.

Often, data can be summarized at several levels. The one you use can affect the patterns you see. In these maps, the high poverty levels shown in the block groups in the upper left (left map) are averaged out when mapped by census tract, as shown in the map on the right.

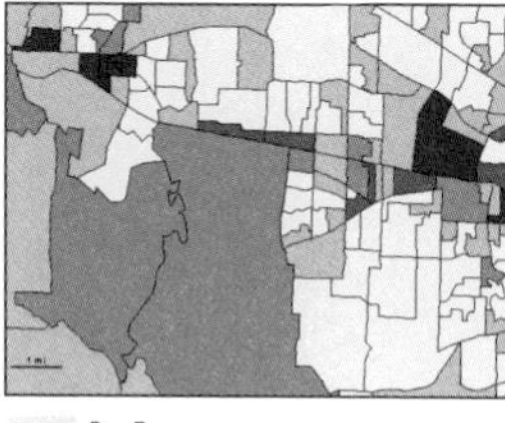

Poverty rate by block group (left) and by census tract. Note that the block groups with a high poverty rate in the upper left are subsumed when the data is aggregated and mapped by census tract.

Keep in mind that data collected for small areas can be summarized up to larger ones, but the reverse is not true. For example, if you know the number of high school students in each block group, you can sum the students for each block group in a tract to get the total for the tract. But if you start with the number of students in each tract, you can't divide them up to each block group in the tract.

To really understand what's going on in a place, you may want to display several maps showing related information. For example, to understand the distribution of the Native American population in a region, you'd want to create maps showing total population by census tract, and the population, percentage, and density of the Native American population.

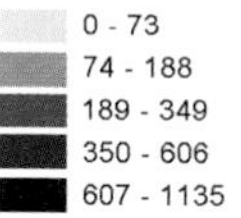

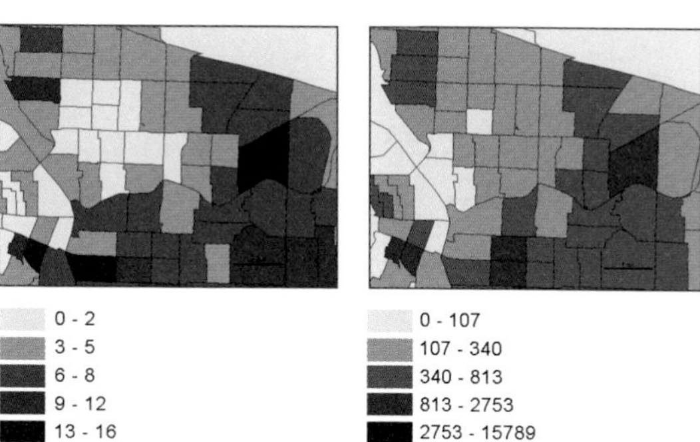

The maps of total population and Native American population (upper left and right) show the numbers of people. The map of percentage Native American (lower left) shows the proportion of the total population, while the density map (lower right) shows the distribution of the Native American population in this area.

Mapping density

Mapping the density of features lets you see the patterns of where things are concentrated. This helps you find areas that require action or meet your criteria, or monitor changing conditions.

In this chapter:

- *Why map density?*
- *Deciding what to map*
- *Two ways of mapping density*
- *Mapping density for defined areas*
- *Creating a density surface*

Mapping density shows you where the highest concentration of features is. Density maps are particularly useful for looking at patterns rather than at the locations of individual features, and for mapping areas of different sizes.

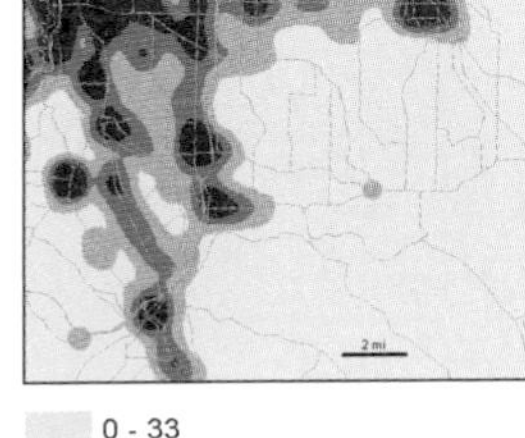

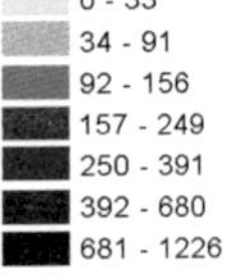

The map on the left shows the locations of businesses. The density surface (on the right) shows the concentration of businesses, rather than their individual locations.

While you can see concentrations by simply mapping the locations of features, in areas with many features it may be difficult to see which areas have a higher concentration than others. A density map lets you measure the number of features using a uniform areal unit—such as hectares or square miles—so you can clearly see the distribution. For example, a crime analyst might map the density of burglaries occurring over a year, per square mile, to compare different parts of the city; a transportation planner might map the density of workers to determine where to place a transit stop so it's nearest the most people.

Mapping density is especially useful when mapping areas, such as census tracts or counties, which vary greatly in size. On maps showing the number of people per census tract, the larger tracts might have more people than smaller ones. But some smaller tracts might have more people per square mile—a higher density.

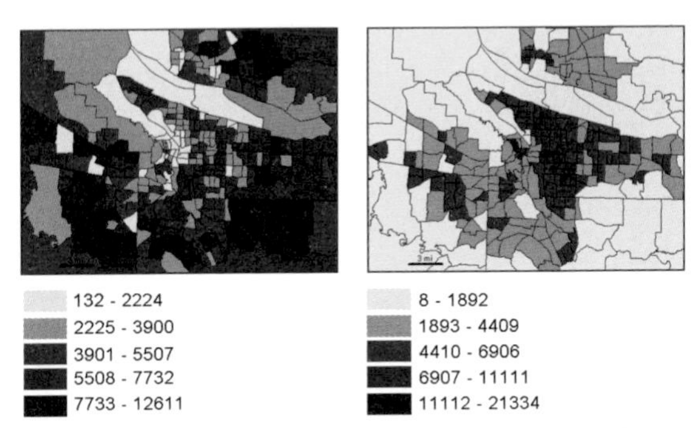

If you want to know approximately how many people each census tract has, map total population (left map). If you want to know where most of the people are concentrated, map population density (right map, showing people per square mile).

MAP GALLERY

Police at the City of Stockton, California, mapped narcotics arrests across the city to see hot spots. The map shows arrests occurring during a one-year period, per square mile. Two hot spots are located near the center of the city, with a third high-density area located to the south. The map helps police commanders assign officers and show city officials and the public where drug activity is concentrated.

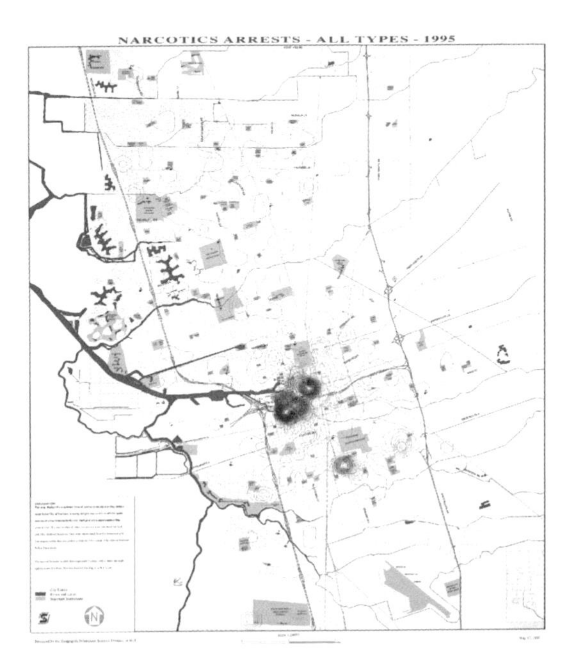

The Southeast Michigan Council of Governments mapped population density for Detroit and surrounding areas as part of a study to find the best location for Secretary of State local service offices. The map, created by calculating the population per square mile for each census tract and shading the tracts accordingly, shows the highest density in red areas. The circles show a 3- and 5-mile radius around each existing service office. Red areas not within a circle are candidates for a new office.

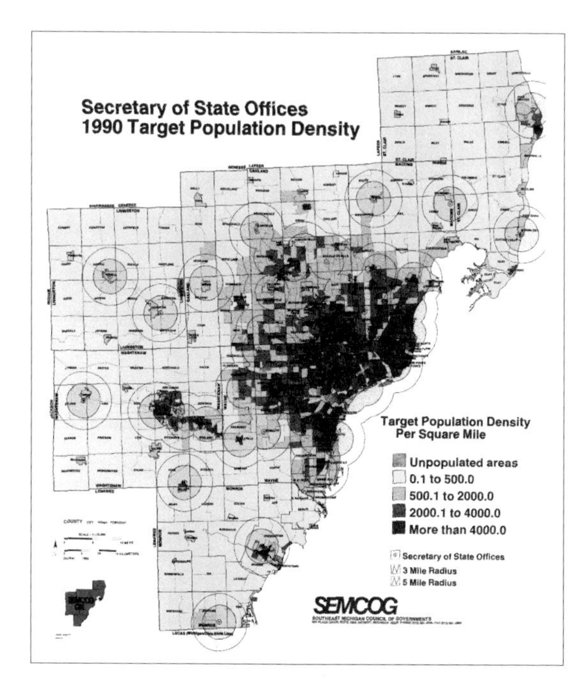

The Duke Power Company mapped densities of fish in Lake Norman, a reservoir in western North Carolina. Hydroacoustic equipment mounted on a boat was used to count the number of fish at regularly spaced locations across the reservoir. The sample data was entered into a GIS and used to create a map showing the number of fish per hectare. Several locations (shown in yellow, orange, and red) have exceptionally high density. Biologists use the information to help monitor where fish are and whether stocks are increasing or decreasing.

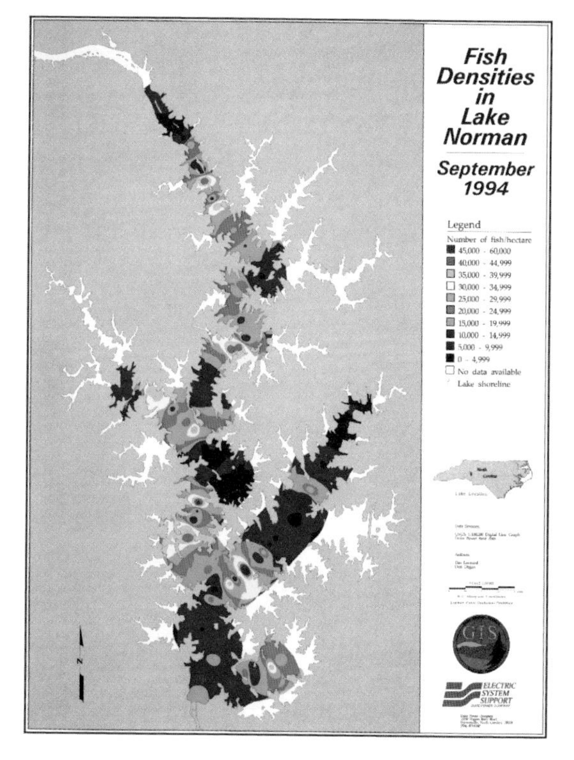

Analysts at the Oregon Department of Forestry used GIS to map cubic meters of timber per hectare produced annually in the state's forests. The map, which shows the highest productivity in the coastal forests (the dark green areas), helps foresters manage the forests.

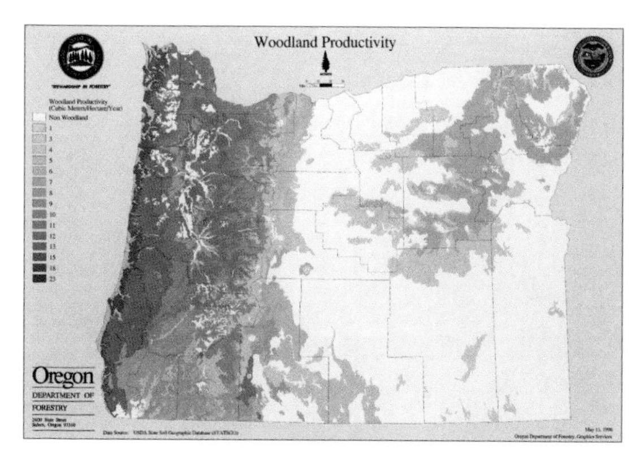

To map density, you can shade defined areas based on a density value, or create a density surface. Before making your map, you should think about the features you're mapping, and the information you need from the map; this will help you decide which method to use.

WHAT KIND OF DATA DO YOU HAVE?

You can use GIS to map the density of points or lines. Usually, these features are mapped using a density surface.

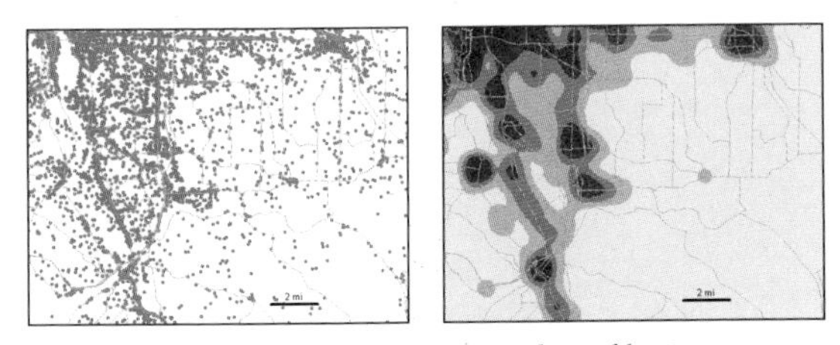

Locations of businesses, and density surface of businesses per square mile

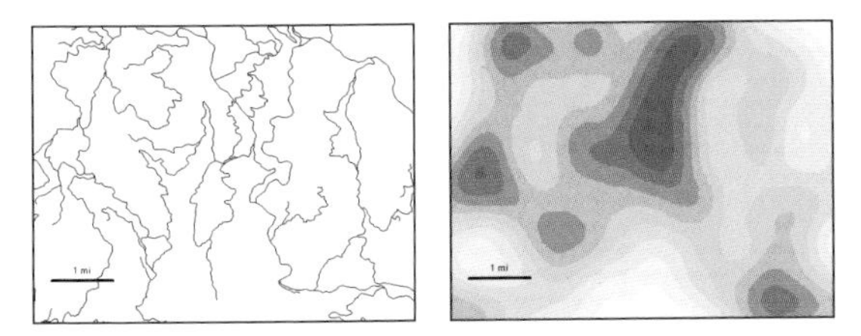

Logging roads, and density of road length (feet per square mile)

Or, you can map data that has already been summarized by defined areas, such as census tracts, counties, forest districts, or other administrative boundaries.

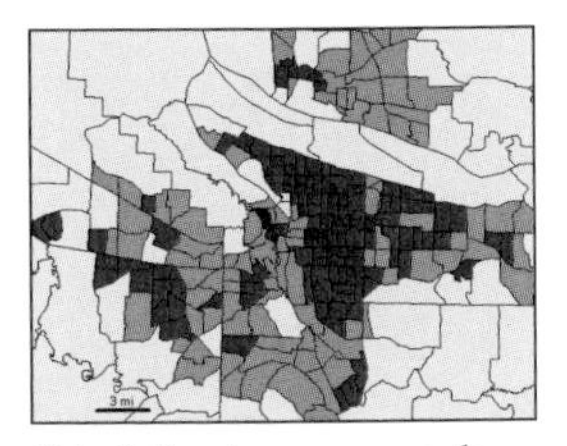

Population per square mile, by census tract

DO YOU WANT TO MAP FEATURES, OR FEATURE VALUES?

You can map either the density of features (for example, the locations of businesses), or of feature values (for example, the number of employees at each business). The resulting patterns can be quite different. The maps below show the density of businesses in an area and the density of workers at those businesses. When employees are mapped, the center of density shifts to the right, between the two concentrations of business locations. To the left is a large number of employees at a single business, creating another center of density. While seeing the density of businesses might be useful for a salesperson making site visits, the density of employees would be useful for a transportation planner wanting to place a transit stop near the largest number of workers.

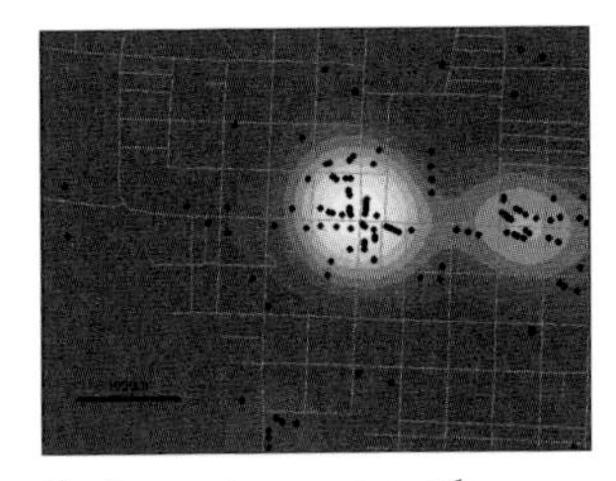

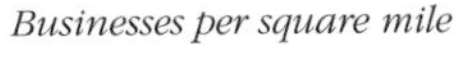

Businesses per square mile

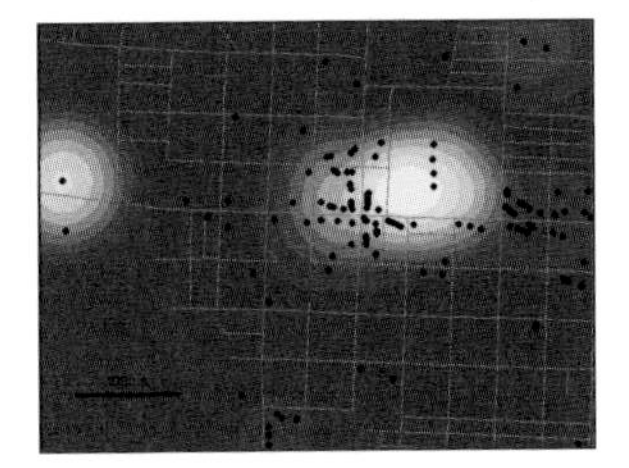

Employees per square mile

Although you can simply map feature locations to see where they're concentrated, creating a density map gives you a measurement of density per area, so you can more accurately compare areas, or know whether certain areas meet your criteria. You can create a density map based on features summarized by defined area, or by creating a density surface.

MAPPING DENSITY FOR DEFINED AREAS

You can map density graphically, using a dot map, or calculate a density value for each area.

You can use a dot map to represent the density of individual locations (people, trees, crimes) summarized by defined areas. Each dot represents a specified number of features, for example 1,000 people, or 10 burglaries. The dots are distributed randomly within each area; they don't represent actual feature locations. The closer together the dots are, the higher the density of features in that area. Dot density maps show density graphically, rather than showing density value. Since you specify the number of features each dot represents, you can use dot density maps to show density when you have many clustered features (rather than mapping each individual feature). Plus, the dots are distributed throughout a defined area, so the map is easier to read.

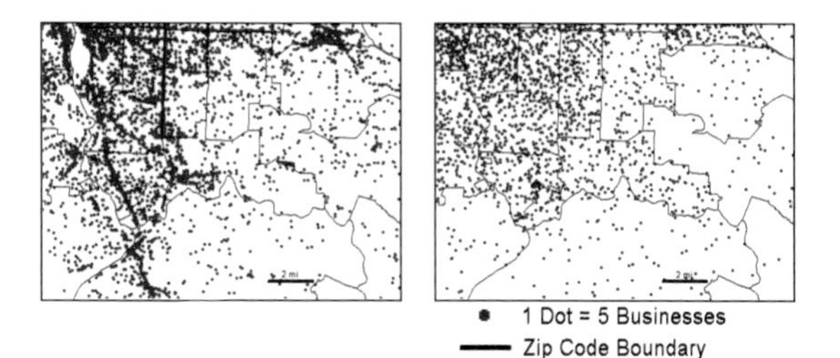

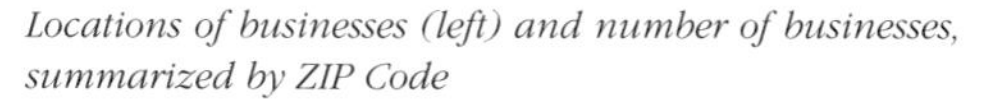

Locations of businesses (left) and number of businesses, summarized by ZIP Code

To calculate a density value for each area, you divide the total number of features, or total value of the features, by the area of the polygon. Each area is then shaded based on its density value. You can see which areas have a higher density, but you will not see the specific centers of density, especially if the areas are large.

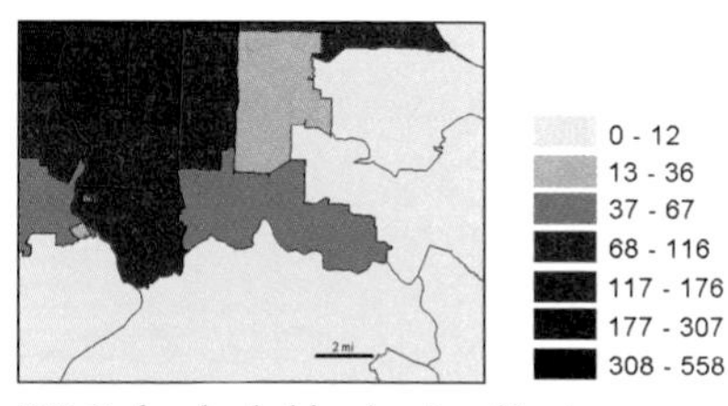

ZIP Codes shaded by density of businesses per square mile

CREATING A DENSITY SURFACE

A density surface is usually created in the GIS as a raster layer. Each cell in the layer gets a density value (such as number of businesses per square mile) based on the number of features within a radius of the cell. This approach provides the most detailed information, but requires more effort.

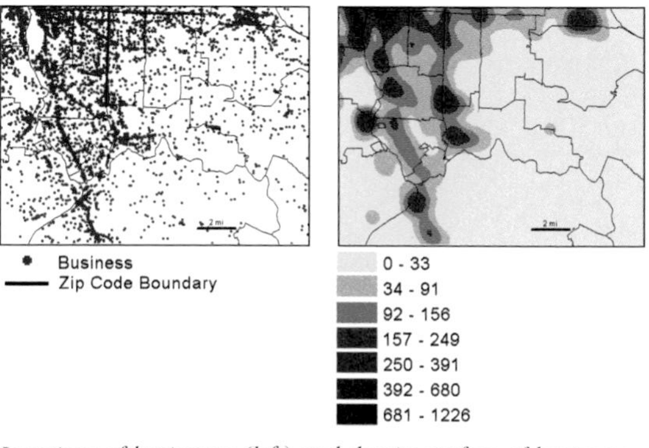

Locations of businesses (left) and density surface of businesses per square mile

You can create a density surface from individual locations, or linear features, such as roads or streams. The point data may be:

- Locations of features, such as customers, crimes, or eagle nests.
- Sample points you've collected data for, such as water quality samples across a lake. These are often regularly spaced, and used to map continuous phenomena.

For lines, the density is usually based on the length per unit area. For example, the total meters of logging roads per hectare.

COMPARING METHODS

Method	Use if...	Output	Trade-offs
Map density by area	You have data already summarized by area, or lines or points you can summarize by area	Shaded fill map or dot density map	Relatively easy, but doesn't pinpoint exact centers of density, especially for large areas; may require some attribute processing
Create a density surface	You have individual locations, sample points, or lines	Shaded density surface or contour map	Gives a more precise view of centers of density, but requires more data processing

CHOOSING A METHOD

Map by defined area if you have data already summarized by area, or you want to compare administrative or natural areas with defined borders, such as census tracts or watersheds.

Create a density surface if you want to see the concentration of point or line features.

Each of these methods is discussed in greater detail in the following pages.

As mentioned earlier, you can map density for defined areas in two ways—you can show density for each area graphically using a dot density map, or you can calculate a density value for each area and shade each based on this value.

CALCULATING A DENSITY VALUE FOR DEFINED AREAS

With this method, you calculate density based on the areal extent of each polygon. First, add a new field to the feature data table to hold the density value. Then, assign the density values by dividing the value you're mapping by the area of the polygon. If the density units are different from the area units, you'll need to use a conversion factor in the calculation to change the area units to the density units. For example, if you're mapping the population density per square mile for each census tract, and the area of each tract is stored in square feet, your calculation would look like this:

```
pop_density =
total_pop / (area / 27878400)
```

since there are 27,878,400 square feet in a square mile. The result would be the population per square mile for each tract.

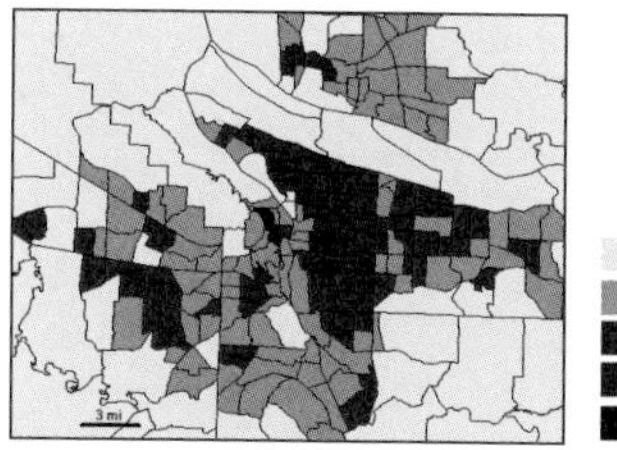

Tract	Area (sq ft)	Population	People per sq mi
406.02	406839193	4791	328
105.00	5345396639	4642	24
427.00	10332982	3442	9287
424.00	18704917	1067	1590
430.00	14282314	1610	3143
333.00	771357260	5974	216

Census tracts by population density (people per square mile)

Density by defined area is usually displayed as a shaded map, using a range of color shades with one or two hues. In this case, density is treated as a ratio and is mapped like any other ratio. Chapter 3, 'Mapping the most and least,' includes a discussion of specifying class ranges and colors for ratios.

Some GIS software, such as ArcInfo and ArcView GIS, lets you also calculate density on the fly—when you create the map, you specify the value you're mapping density for and the attribute containing the area of each feature. The GIS calculates the density values and shades each area accordingly. The density values aren't stored in the database—they're only temporary.

Keep in mind that the density value for each polygon applies to the entire polygon; the actual density at any given location within the polygon may vary greatly from this value. This is especially true for large polygons. In this map, some larger ZIP Codes with very low overall density of businesses actually contain areas with a very high density of businesses.

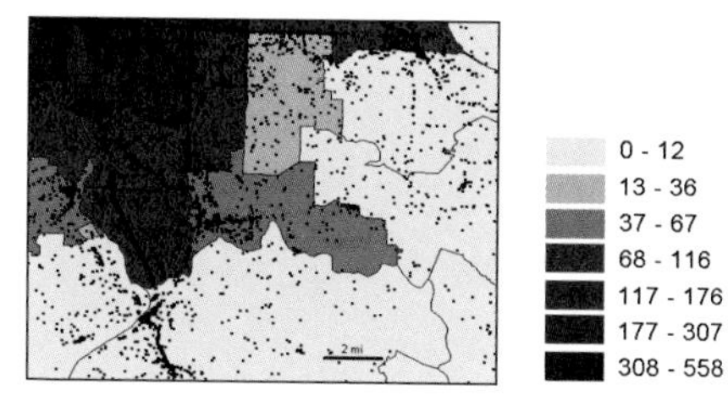

Density of businesses per ZIP Code, with business locations

CREATING A DOT DENSITY MAP

With this method, you map each area based on a total count or amount, and specify how much each dot represents. The GIS divides the value of the polygon by the amount represented by a dot to find out how many dots to draw in each area. If, for example, each dot represents 200 people, and a census tract has 6,000 people in it, the GIS will draw 30 dots in that tract. The GIS places dots randomly within the area; the dots don't represent the actual locations of features.

Dot maps give the reader a quick sense of density in a place. For example, two census tracts of different sizes, but the same total population, would appear the same color on a shaded map, but a dot density map would show that the smaller tract has a higher density, since there would be the same number of dots in a smaller space.

Same total population

132 - 2224
2225 - 3900
3901 - 5507
5508 - 7732
7733 - 12611

Different density

1 Dot = 200 People
Census Tract Boundary

The two tracts have roughly the same total population and are shaded with the same color (upper map); the dot density map (lower) shows the difference in density between them.

A dot map simply represents density graphically. The dots represent total numbers or values in each area, rather than a calculated density value.

When creating a dot density map, you specify how many features each dot represents, and how big the dots are. You may need to try several combinations of amount and size to see which one best shows the patterns.

The larger the amount represented by each dot, the more spread out they will be. Select a value that ensures the dots are not so close as to form solid areas that obscure the patterns, or so far apart as to make the variations in density hard to see.

You can change the dot size to emphasize the patterns. Make sure the dots are not so large that areas with high density obscure the patterns.

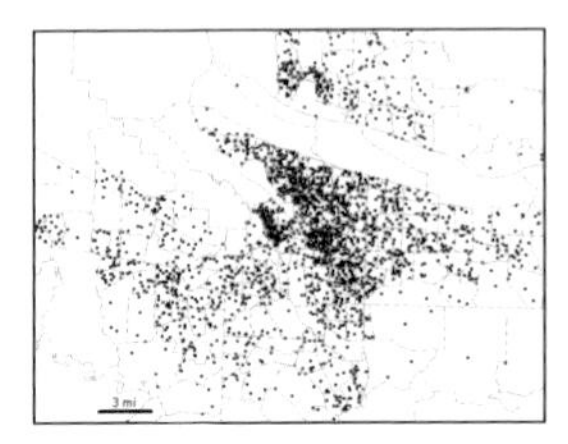

Amount and size are appropriate; dots show patterns.

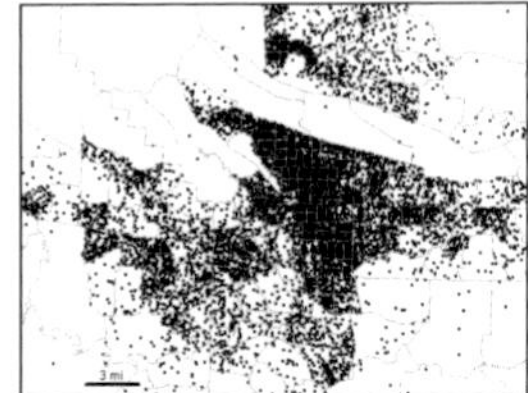

Amount too small; dots obscure patterns.

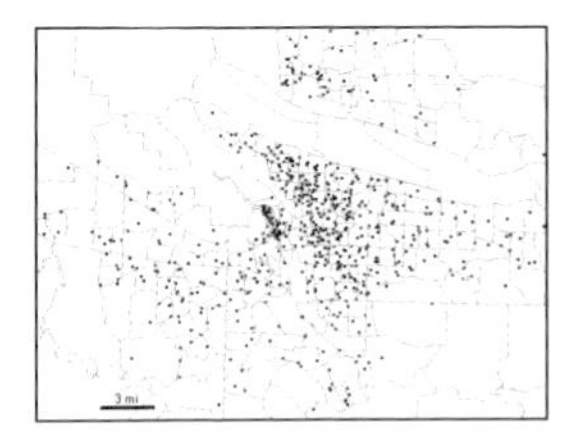

Amount too large; dots are too far apart to show patterns.

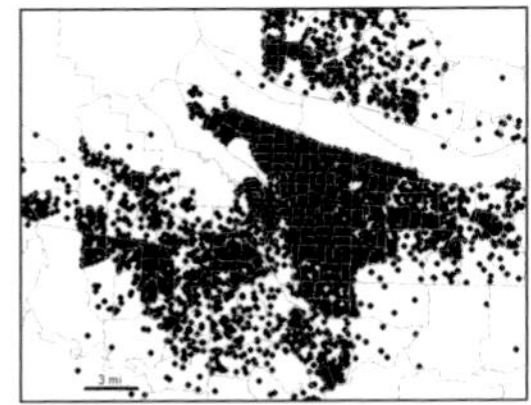

Size too large; dots obscure patterns.

The areas you choose to map may affect the patterns. For example, the density when mapped by census tract may look quite different than when mapped by county.

Households by county (1 dot = 500 households).

Households by tract, with county boundaries (1 dot = 500 households).

When creating a dot density map, you often display the dots based on smaller areas, but draw the boundaries of larger areas. That way, the boundaries won't obscure the dots. The map presents a more realistic view of density than the one using dots based on counties. You should state on the map the actual areas used for summarizing the data.

Households by census tract (1 dot = 500 households).

Households by tract, with county boundaries (1 dot = 500 households).

If you have individual features but want to map density summarized by defined areas...

You can use GIS to summarize features or feature values for each polygon. You might have to do this to compare areas to find out which meet your criteria, or if you're creating a series of maps for the areas, such as census tracts by total population, percentage of the population aged 18 to 30 years, and density of crimes. It would be easier to see relationships if crimes were mapped by tract as well, rather than as individual locations. You do this by tagging each location with the ID of the area it falls inside, totaling the features for each area, and dividing the total features by the size of the area (see chapter 5, 'Finding what's inside').

Density surfaces are created in a GIS as raster layers—the GIS calculates a density value for each cell in the layer. Density surfaces are good for showing where point or line features are concentrated.

WHAT THE GIS DOES

To create a density surface, the GIS defines a neighborhood (based on a search radius you specify) around each cell center. It then totals the number of features that fall within that neighborhood and divides that number by the area of the neighborhood. That value is assigned to the cell. The GIS moves on to the next cell and does the same thing. This creates a running average of features per area, to create a smoothed surface.

If you are using a data value instead of the number of features, the GIS totals that value for all features in the neighborhood and divides it by the area of the neighborhood. So, if you're mapping the density of employees based on the number at each business, the GIS would identify the businesses in the neighborhood, sum the number of employees at those businesses, and divide that number by the area of the neighborhood.

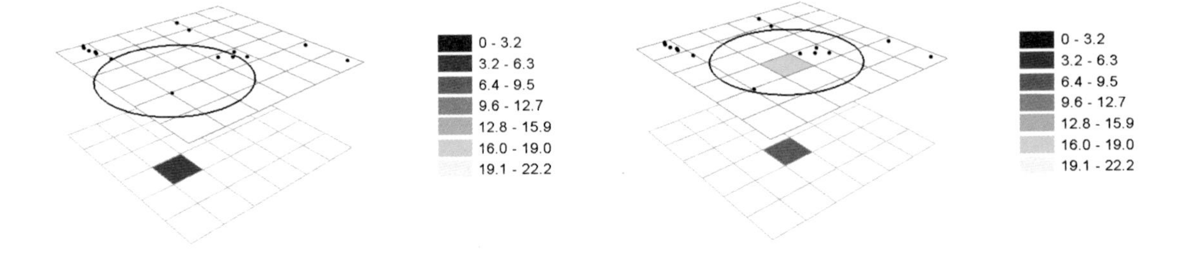

With a search radius of 100 feet, the area of the search neighborhood is 31,349 square feet, or 0.72 acres. This cell has three businesses within its search neighborhood, so its density value is 3 divided by 0.72, or 4.2 businesses per acre. It is assigned to the second lowest class.

The next cell has five businesses within its neighborhood—the same three the previous cell had, plus two additional ones. 5 divided by 0.72 = 6.9, so this cell gets assigned to the next higher class.

CALCULATING DENSITY VALUES

Several parameters that you specify affect how the GIS calculates the density surface, and thus what the patterns will look like. These include cell size, search radius, calculation method, and units.

Cell size

The cell size determines how coarse or fine the patterns will appear. The smaller the cell size, the smoother the surface. However, since there will be more cells than when using a larger cell size, the resulting layer will take longer to process and will require more storage space. A larger cell size will process faster but result in a coarser surface. If each cell is so large as to include many features, subtle patterns may be obscured.

Since cells are square, the cell size is specified as the length of one of its sides. In general, set the cell size so you have between 10 and 100 cells per density unit. For example, if you're mapping population per square kilometer, and the cell size is in meters, your cell size should be between about 100 meters and 300 meters. To calculate cell size, first convert the density units (square kilometers) to cell units (meters), then divide by the number of cells per density unit to get the area of each cell. Since cell size is specified by the length of one side of a cell, you take the square root of the cell area to get this. For example, using 100 cells per density unit:

1 Convert density units to cell units

```
1 sq. km = 1,000 m * 1,000 m
= 1,000,000 sq. m
```

2 Divide by the number of cells

```
1,000,000 sq. meters / 100 cells
= 10,000 sq. meters per cell
```

3 Take the square root to get the cell size (one side)

```
√10,000 m = 100 m
```

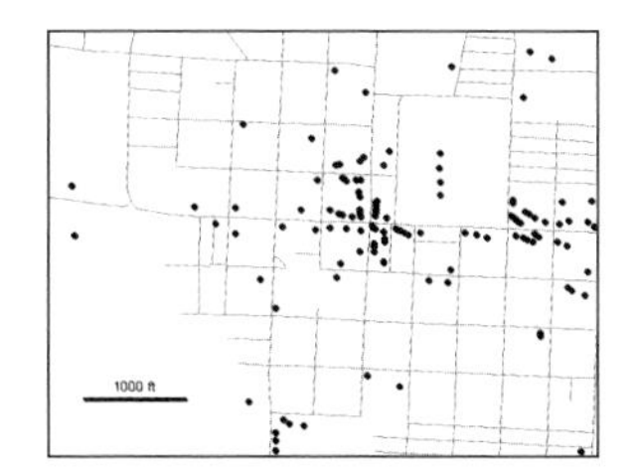

Business locations

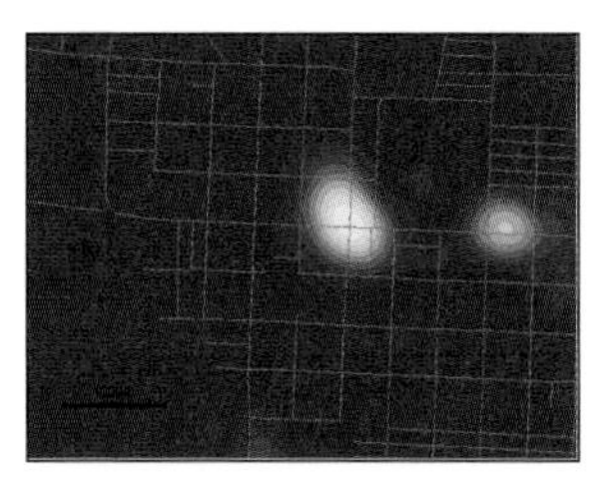

Cell size = 5 feet; smooth surface, but takes longer to process.

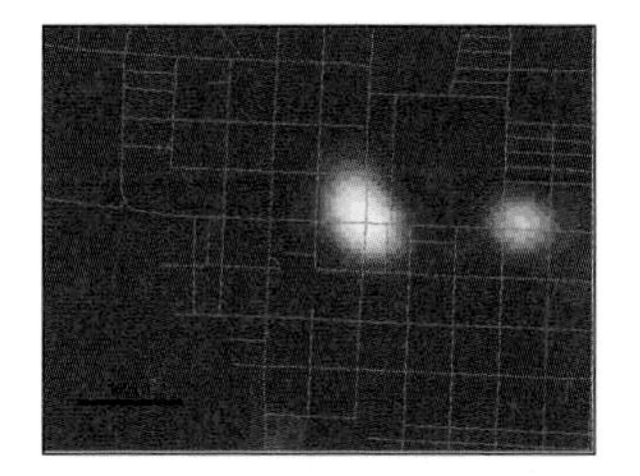

Cell size = 50 feet; shows the same patterns as when using a smaller cell size, but processes faster and takes up less storage space.

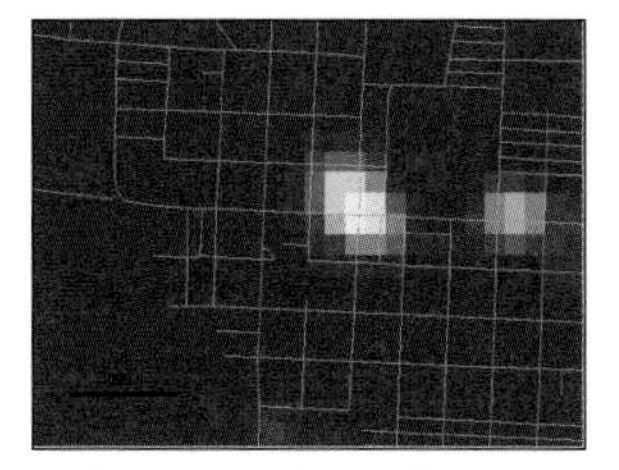

Cell size = 200 feet; cell size is too large, so detail of patterns starts to disappear.

Search radius

Generally, the larger the search radius, the more generalized the patterns in the density surface will be. With a larger search radius, the GIS considers more features when calculating the value of each cell. The number of features (or the total of the feature data values) is divided by a correspondingly larger area. A smaller search radius usually shows more local variation. However, if the search radius is so small that most cells have very low density values, broader patterns in the data may not show up.

The search radius units and the density units do not have to be the same; you can calculate density per square mile, but specify search radius in feet.

Density of businesses per square mile

Search radius = 20 feet; patterns are difficult to see.

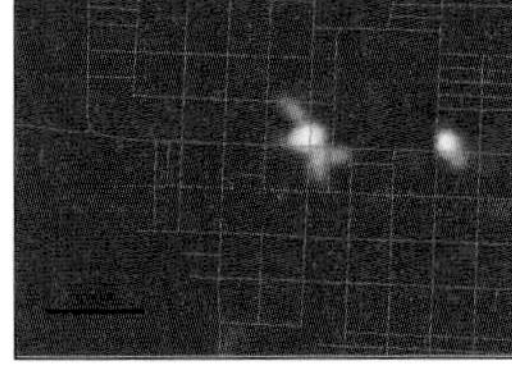

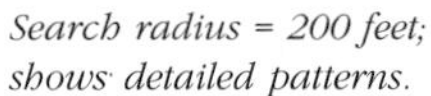

Search radius = 200 feet; shows detailed patterns.

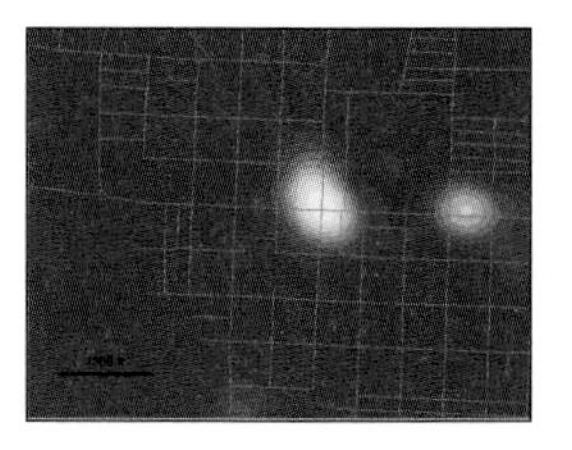

Search radius = 500 feet; patterns become generalized.

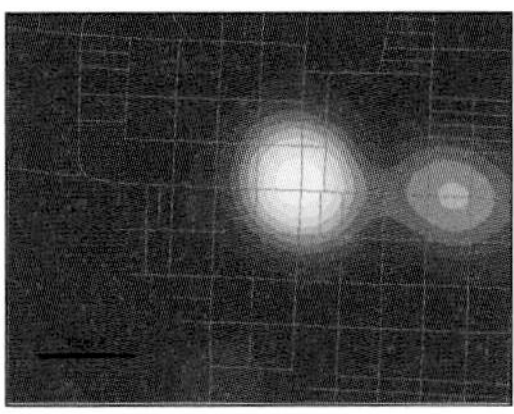

Search radius = 1,000 feet; patterns are too generalized—detail disappears.

Calculation method

The GIS uses one of two methods for calculating the cell values. The simple method counts only those features within the search radius of each cell. The result is a series of rings—one around each cell—that overlap each other. Cells with no features within the search radius are not assigned a value.

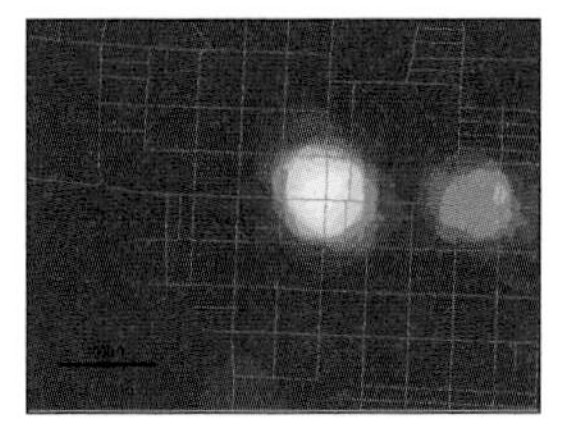

Simple calculation

The weighted method uses a mathematical function to give more importance to features closer to the center of the cell. The weighting drops off rapidly for features beyond the search radius. Every cell in the layer is counted and assigned a value (albeit a very small value for cells that are distant from any features). The result is a smoother, more generalized density surface. In many cases, using the weighted method produces a map with patterns that are easier to interpret.

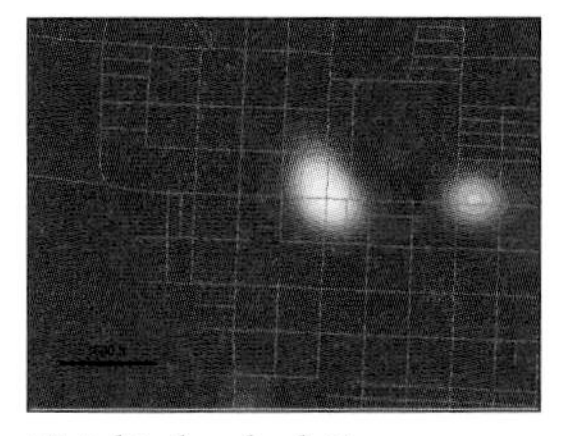

Weighted calculation

Units

The GIS lets you specify the areal units in which you want the density values calculated. You should choose a value for units that reflects the features you're mapping. For example, using square meters might be appropriate for mapping plants or insects, since you can have several of these per square meter, while hectares or square miles would be appropriate for businesses or people.

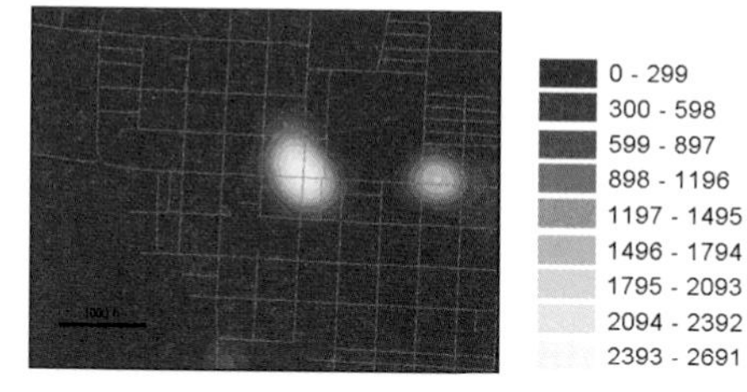

Number of businesses per square mile

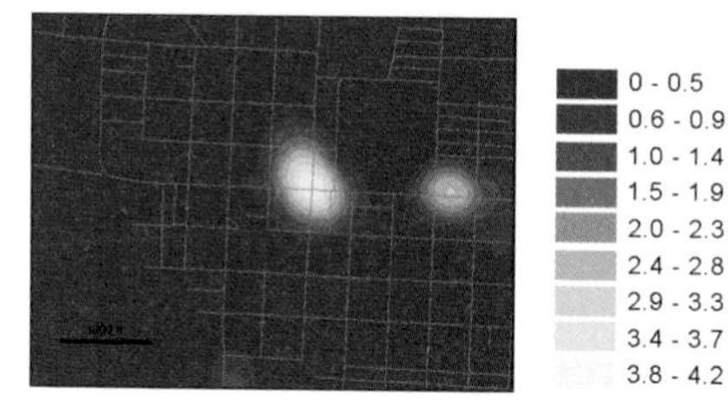

Number of businesses per acre

If the areal units are different from the cell units, the values in the legend are extrapolated—there may not actually be that many features in the mapped area. If the areal units are square miles, the values show how many businesses you'd need within a square mile to have that density. In this example, the mapped area is about 1 square mile and contains about 130 businesses. The map using square miles, while valid as a measure of density, may imply that there are more businesses than there actually are. Using an areal unit of acres gives more realistic values, and doesn't change the patterns on the map.

If you have data summarized by defined areas but want to create a density surface...

You can use the center points, or "centroids," of defined areas, such as ZIP Codes, to create the density surface, based on the value assigned to each area. This works best if you have a large number of points that are fairly evenly distributed. For example, you could use census tract centroids to create a density surface of population. You might do this to highlight the patterns in the map and place less emphasis on the individual polygons.

Census tracts shaded by total population

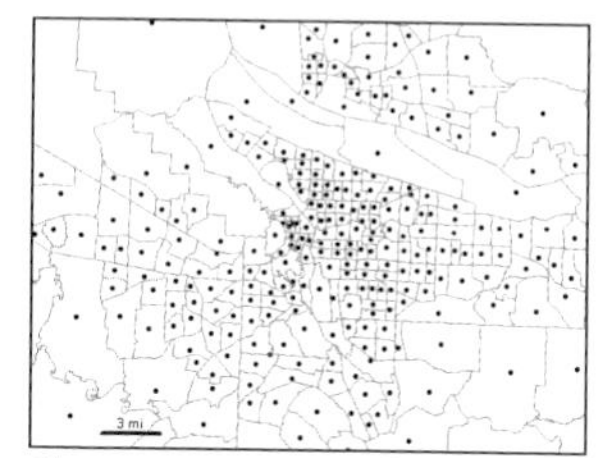

Tract	Population
402.00	8526
403.00	3583
404.01	8251
405.01	8651
404.02	10302
335.00	2577

Census tract centroids, with attached population values

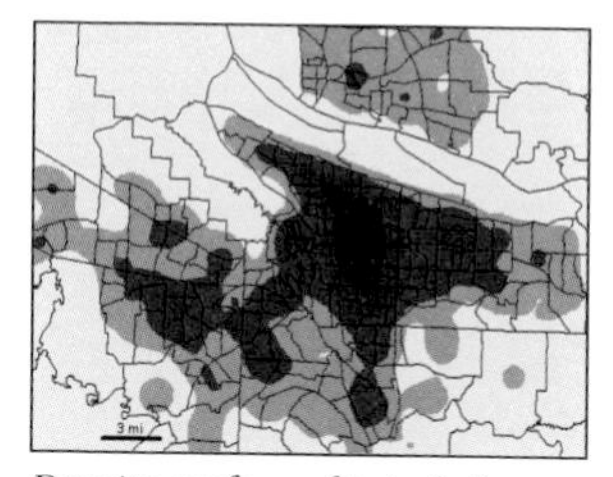

Density surface of population created from census tract centroids, with census tract boundaries for comparison

DISPLAYING A DENSITY SURFACE

You can display a density surface using graduated colors, or using contours.

Using graduated colors

As with other surface layers, since each cell could have a unique density value, you'll need to classify the values to see the patterns.

You can create custom class ranges (specifying the high and low value for each class yourself), or specify a standard classification scheme and let the GIS create the classes for you. Here are the most common classification schemes:

Natural breaks—Class ranges are based on groupings of data values. In the example shown on the right, this classification highlights the areas of highest density, while revealing subtle patterns.

Quantile—Each class has the same number of cells in it. In the example, more cells are forced into the higher classes, obscuring the highest centers of density.

Equal interval—The difference between the high and low value is the same for each class. In the example, the areas of highest density are highlighted, at the expense of subtle patterns in the data.

Standard deviation—The classes are defined by a number of standard deviations from the mean of all values in the layer. This scheme highlights the areas of particularly high or low values.

Chapter 3, 'Mapping the most and least,' includes a detailed discussion of creating classes, and classification methods.

Employees per square mile

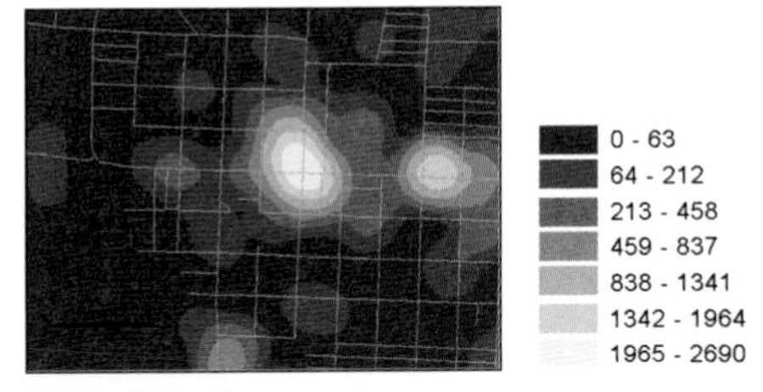

Natural breaks

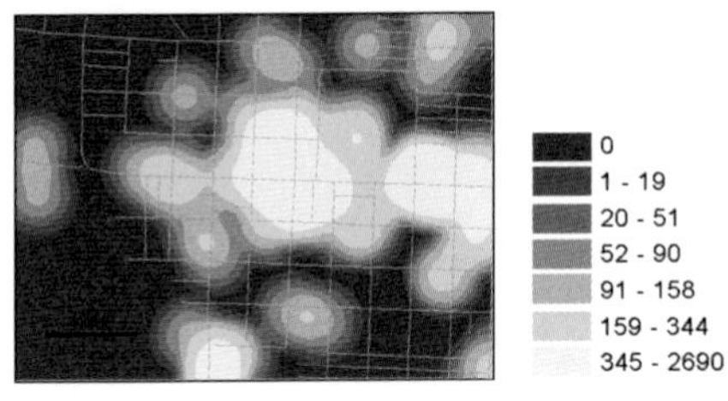

Quantile

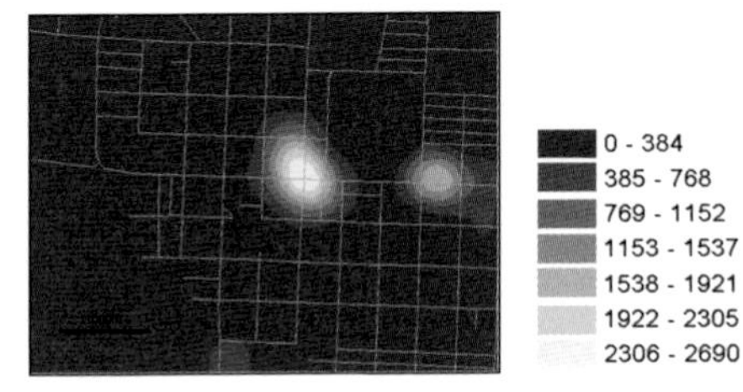

Equal interval

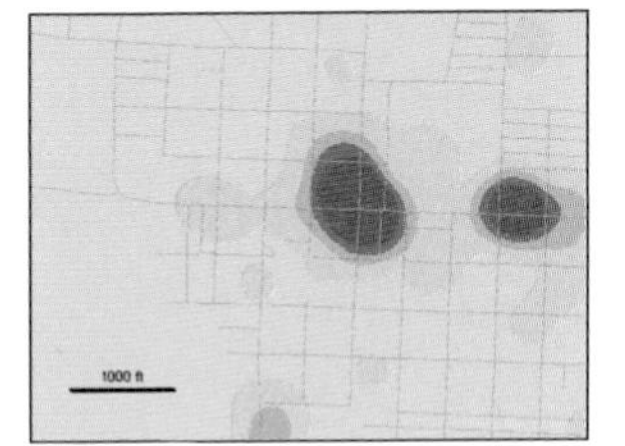
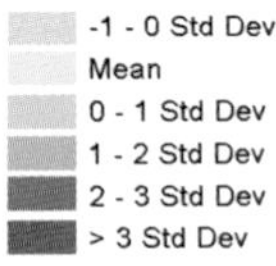

Standard deviation

The GIS lets you specify the number of classes to which to assign the density values. More classes create a smoother looking map. Using too many classes, though—more than 15 or so—doesn't add information to the map since the colors begin to blend, making the classes hard to distinguish. Using a few classes (fewer than three or four) highlights the areas with the highest density, but may not show subtleties in the patterns.

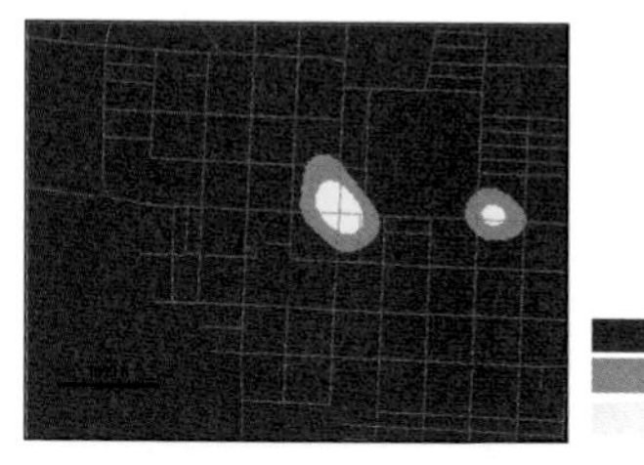

Too few classes

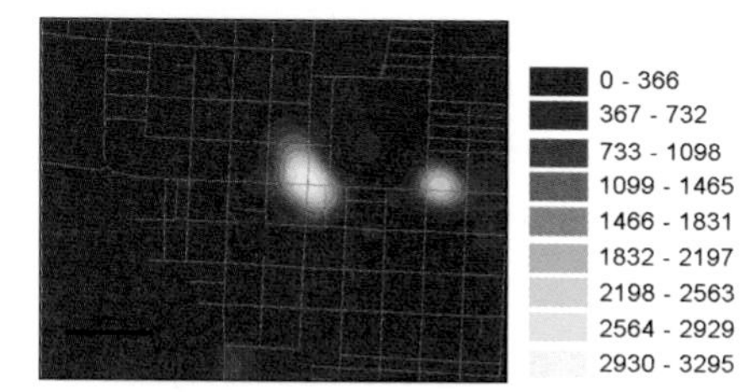

Good number of classes

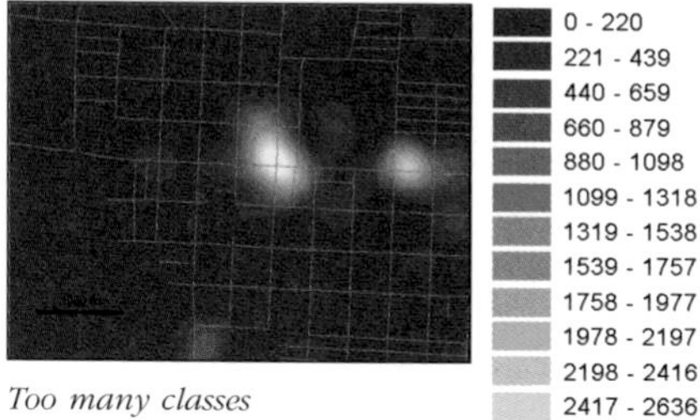

Too many classes

Businesses per square mile. More classes can make subtleties in the patterns clearer—up to a point.

Density surfaces are usually displayed using shades of a single color. If you're using a standard deviation classification, use shades of one color for values below the mean, and shades of another for values above the mean.

Typically, areas of higher value are shown using darker colors, since most people equate darker with "more." However, you can create an effective map by using lighter colors for higher density; the reader's eye is drawn to those areas, while the areas of low density recede.

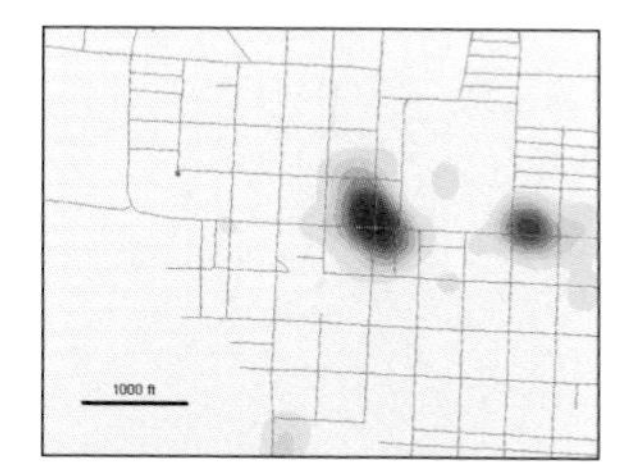

Darker = higher density

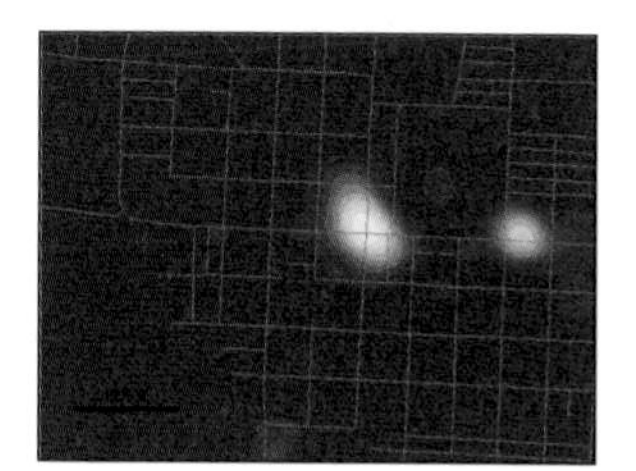

Lighter = higher density

Using contours

Contour lines connect points of equal density value on the surface. Most GIS software, including ArcInfo and ArcView GIS, will create contours automatically from a surface. You simply specify the contour interval—that is, how far apart in value the contour lines are.

Contours are good for showing the rate of change across the surface—the closer the contours, the more rapid the change.

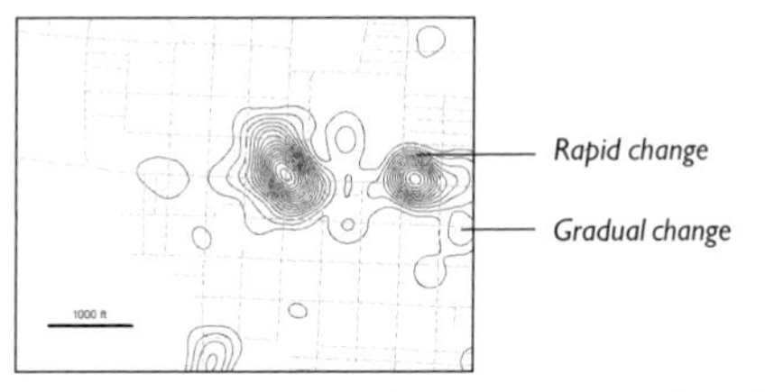

Businesses per square mile; Contour interval = 200

Choose an interval that will show the patterns in areas of little variation without becoming too close together in areas of high variation. There is no hard and fast rule, since it depends on how much and how quickly values change across the surface.

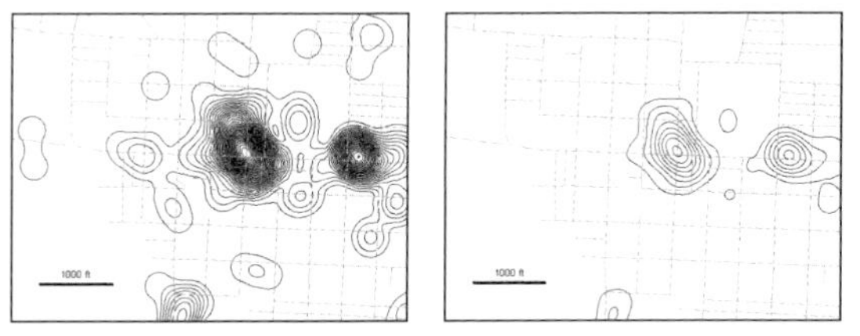

In the map on the left, subtle patterns are apparent, but the contours are hard to distinguish in areas of high density. In the map on the right, contours are distinct, but some information is lost.

Combining contours with a shaded density surface lets the reader quickly see the areas of highest density as well as the rate of change.

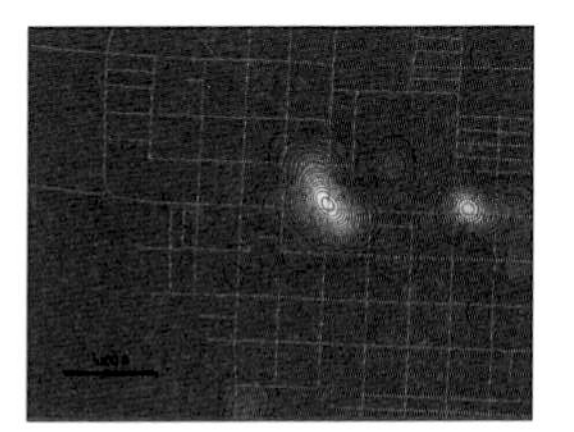

LOOKING AT THE RESULTS

The patterns on your map partially depend on how you created the density surface. For example, the two maps below show the density of logging roads in a forested area, based on the length of road per acre. The one on the left was created using a small search radius, and shows several centers of density. The map on the right, using a larger search radius, shows a single area of high density. A biologist looking for undisturbed sites for a wildlife study would use a search radius based on assumptions about the impact of logging roads on the particular species. A species more sensitive to the presence of roads would require using a larger search radius.

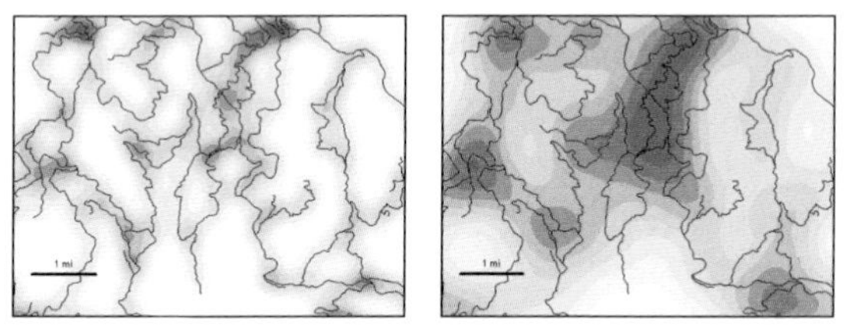

Logging road density. The map on the left uses a smaller search radius and shows more local variation; the map on the right uses a larger search radius and shows a more generalized view. The darker colors indicate higher density.

A density surface can show how values vary across a region. This map shows land value per square foot. The right half of the map shows little variation in land values, while the left side shows more variation. That might indicate a greater variety of land uses in the area with a higher variation in density.

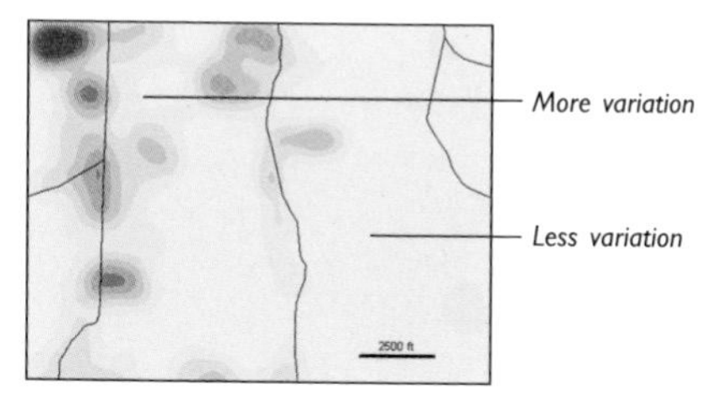

Density surface of land value per square foot. Major roads are also shown.

The patterns in a density surface are affected by the distribution of sample points. The more sample points, and the more dispersed they are, the more valid the patterns will be. Be aware that the values in the areas between the points are estimates.

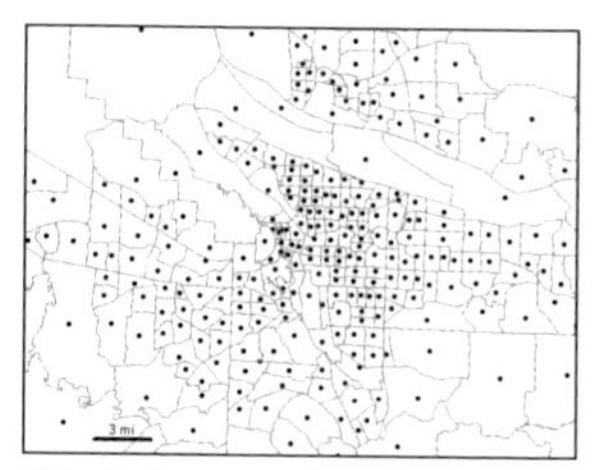

Population sample points (census tract centroids)

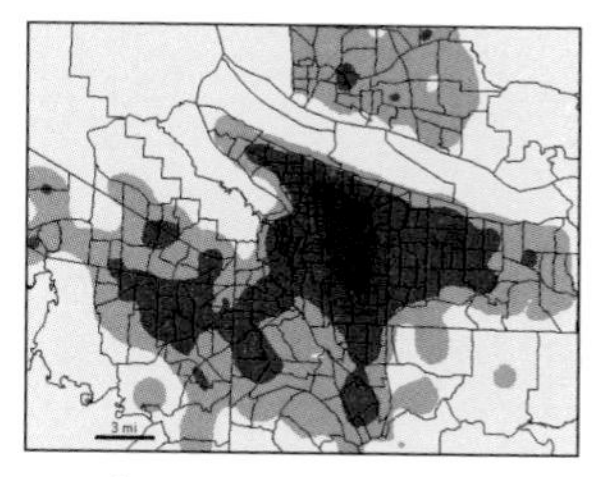

Population density based on census tract centroids. Values in areas having many well-spaced input points are likely to be more valid than values in areas with few sample points.

Also keep in mind that there may not actually be any features where the highest density is, since the GIS calculates values based on a neighborhood around each cell. On this map of the density of employees, there are no businesses, and thus no employees, located in the cells with highest density values (the light area on the right side of the map). In this case, the highest density falls between several major employers. For the area on the left, however, since there are no other major employers nearby, the highest density is at the location of a business.

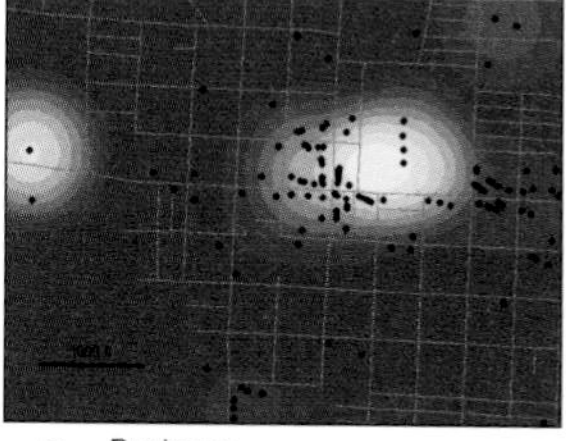

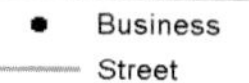

Locations of businesses with a density surface of employees at those businesses (employees per square mile)

When creating a density surface, the interpolation process generalizes and smooths the data, so that extreme high and low values may disappear. This makes the patterns easy to see, but to get a better picture of what's going on in a place, you should also map the locations of features from which the density surface was calculated—either with the density surface, or on a separate map.

Finding what's inside

Finding what's inside lets you see whether an activity occurs inside an area, or summarize information for each of several areas so you can compare them.

In this chapter:

- *Why map what's inside?*
- *Defining your analysis*
- *Three ways of finding what's inside*
- *Drawing areas and features*
- *Selecting features inside an area*
- *Overlaying areas and features*

People map what's inside an area to monitor what's occurring inside it, or to compare several areas based on what's inside each.

By monitoring what's going on in an area, people know whether to take action. For example, a district attorney would monitor drug-related arrests to find out if an arrest is within 1,000 feet of a school—if so, stiffer penalties apply. A fire chief mapping the area covered by a toxic plume from a chemical spill would want to see which evacuation areas are outside the plume area to know where to send evacuees.

Summarizing what's inside each of several areas lets people compare areas to see where there's more and less of something. For example, a police chief might want to create a monthly map of burglaries in each reporting district to see hot spots. A conservation organization might want to find out which watersheds have the most remaining ancient forest, to target areas for conservation.

MAP GALLERY

Forest products company Boise Cascade mapped the amount of each habitat type inside each subwatershed in the Gold Fork Watershed of Idaho so it can efficiently manage the forests within each watershed.

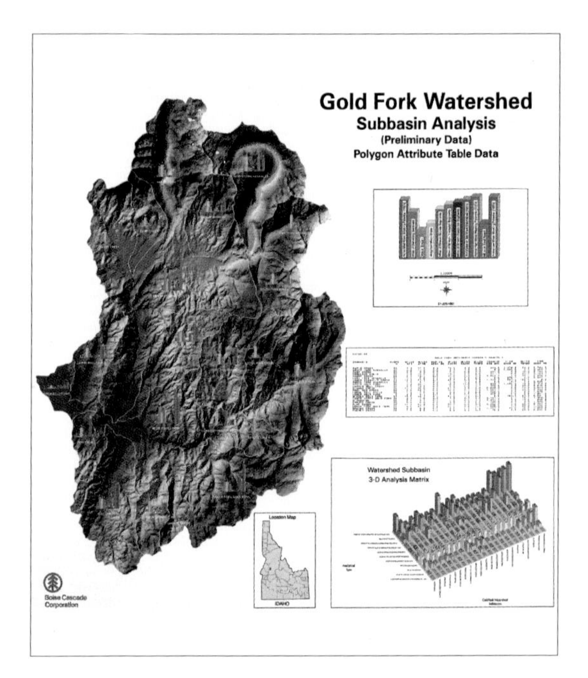

Engineering firm Parson Brinckerhoff created a series of maps showing alternative routes for a proposed light rail line. Each map showed hazardous materials sites inside a 1,000-foot buffer around the rail line, along with the depth to the water table. The maps were used to analyze potential impacts to water resources along the route.

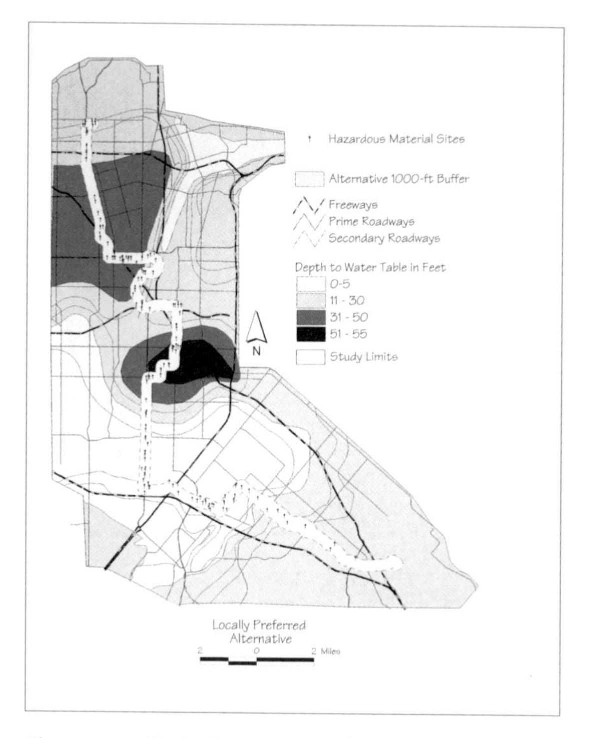

Planners at Clark County, Nevada, created a map showing noise contours, along with streets and property boundaries, around Nellis Air Force Base. The map helps them ensure schools and hospitals are not built in the areas with the highest noise levels.

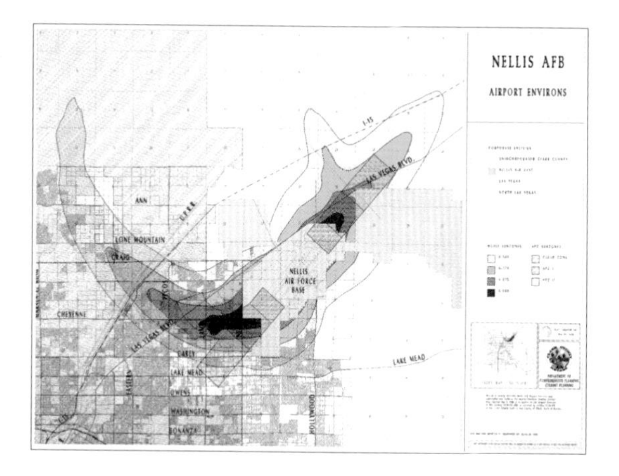

Conservation International, working with local agencies in Madagascar, created a map of protected areas and habitats for the entire country. By calculating the percentage of each type of habitat inside protected areas, government agencies are able to see which are least protected.

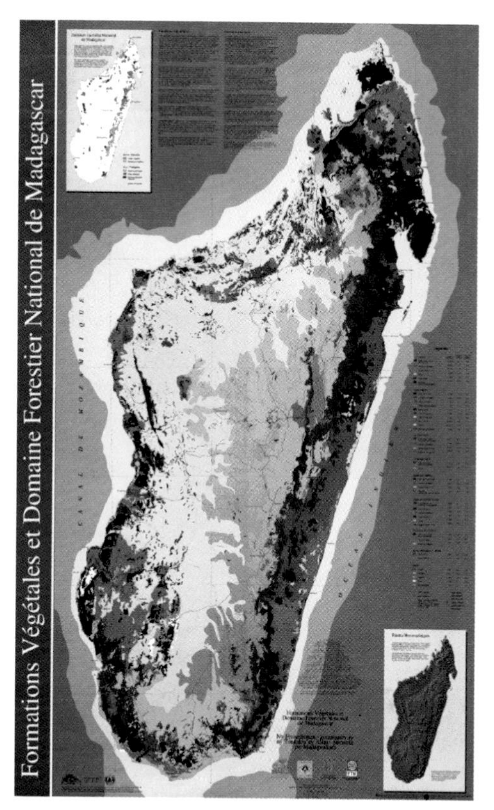

To find what's inside, you can draw an area boundary on top of the features, use an area boundary to select the features inside and list or summarize them, or combine the area boundary and features to create summary data.

The method you use depends on the data you have and the information you need from the analysis.

YOUR DATA

You need to consider how many areas you have, and what type of features are inside the areas.

Are you finding what's inside a single area or each of several areas?

You can find what's inside a single area, or inside each of several areas.

Single area

Finding what's inside a single area lets you monitor activity or summarize information about the area. Single areas include:

- A service area around a central facility, such as a library district or fire response area

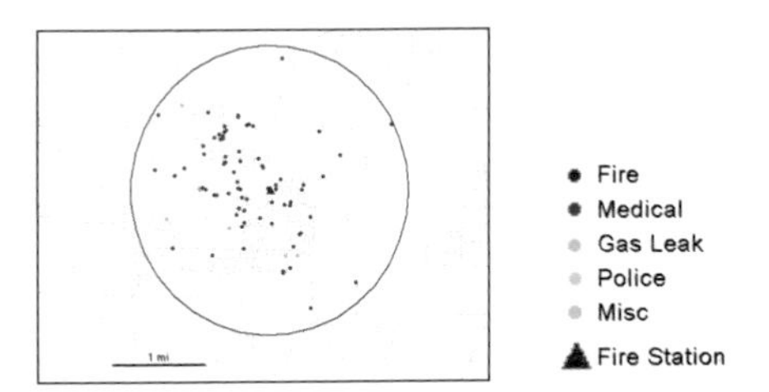

Calls to 911 within a mile-and-a-half of a fire station

- A buffer that defines a distance around some feature, such as a stream buffer that's off limits to logging

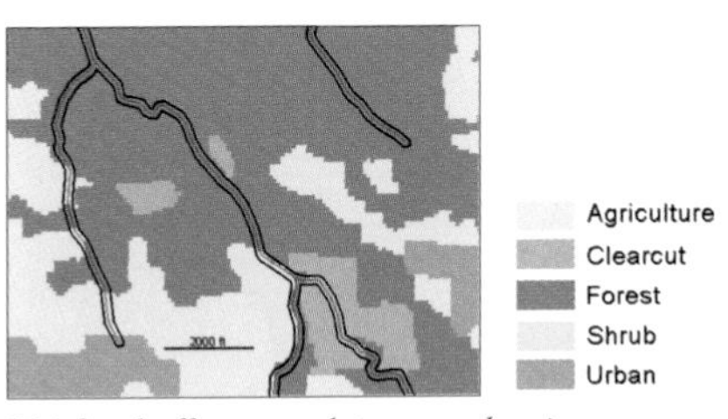

200-foot buffer around streams, showing area off-limits to logging

- An administrative or natural boundary, such as a police beat, parcel of land, or watershed

Soil types inside a land parcel

- An area you draw manually, such as a proposed sales territory

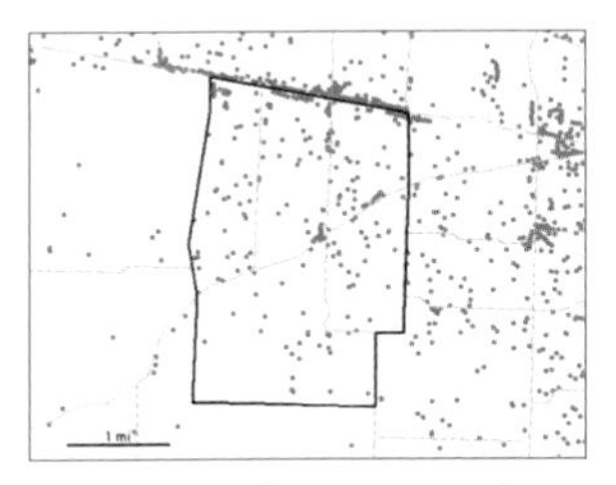

Customers within a proposed sales territory

- The result of a model, for example, the boundaries of a floodplain modeled in a GIS

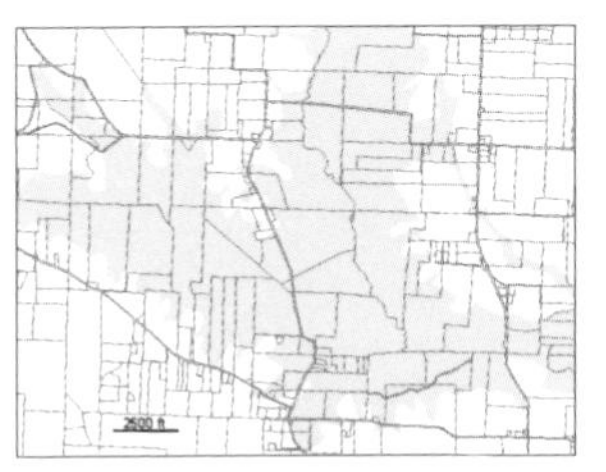

Parcels within a 100-year floodplain

You can also find what's inside several areas you're treating as one—for example, the number of businesses within a group of ZIP Codes.

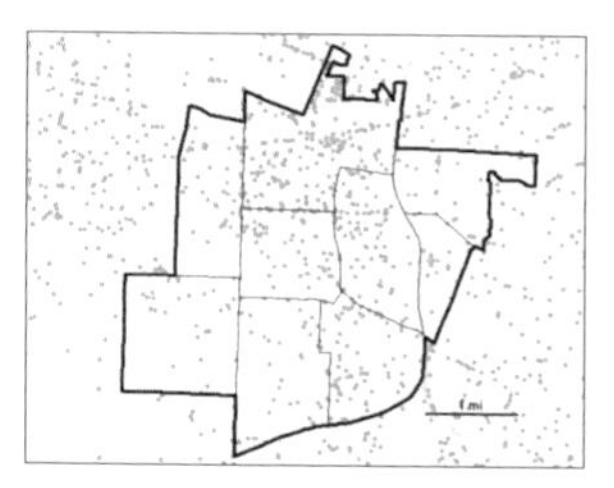

Businesses inside a group of ZIP Codes

Multiple areas

Finding how much of something is inside each of several areas lets you compare the areas. The areas can be:

- Contiguous, such as ZIP Codes or watersheds

ZIP Codes are contiguous.

- Disjunct, such as state parks

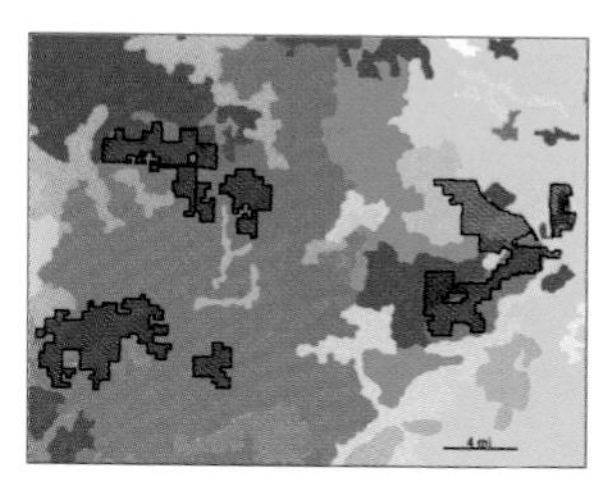

These state park boundaries are disjunct.

- Nested, such as 50- and 100-year floodplains, or the area within 1, 2, and 3 miles of a store

You'll want to be able to identify each area uniquely using a name—for example, a named watershed—or a unique number, such as a census tract number.

Are the features inside discrete or continuous?

Discrete features are unique, identifiable features. You can list or count them, or summarize a numeric attribute associated with them. They are either locations, such as student addresses, crimes, or eagle nests; linear features, such as streams, pipelines, or roads; or discrete areas, such as parcels.

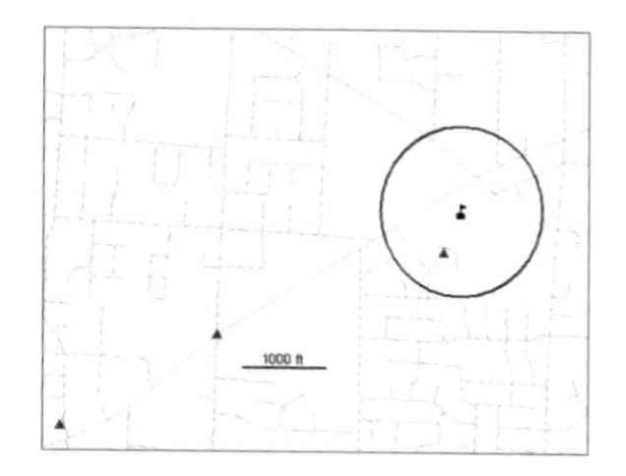

Locations—drug-related arrests near a school

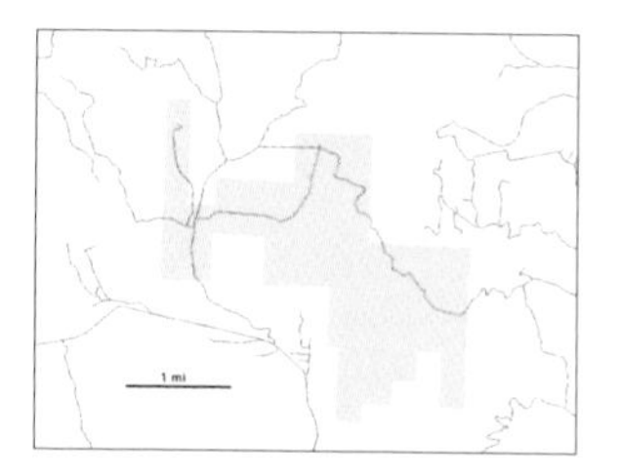

Linear features—roads through a protected area

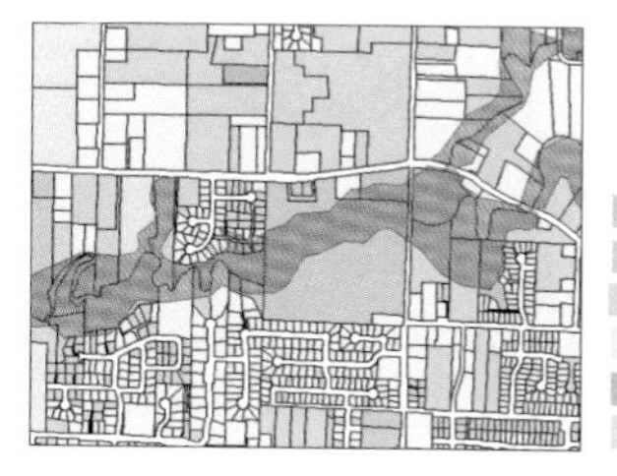

Discrete areas—parcels within a floodplain

Continuous features represent seamless geographic phenomena. You can summarize the features for each area. Continuous features include:

- Spatially continuous categories or classes, such as vegetation type or elevation range. You can find out how much of each category or class occurs inside each area—for example, the amount of each vegetation type in each watershed.

Soil types are represented by continuous categories.

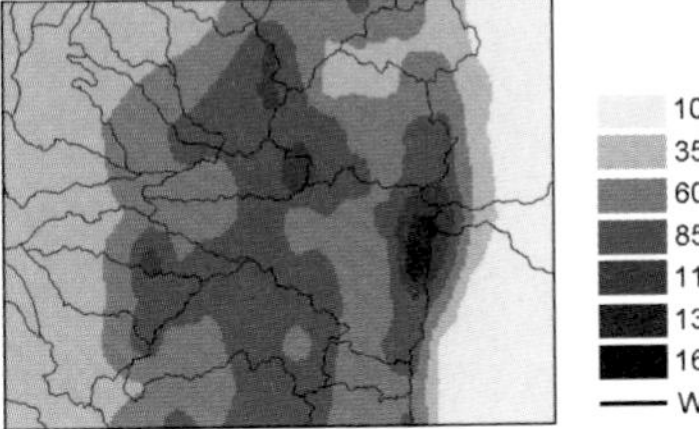

Bands of precipitation (inches per year) are represented by continuous classes.

- Continuous values. These are numeric values that vary continuously across a surface. They can be measures, such as temperature, elevation, or precipitation. They can also be values that have been derived from other data, using the GIS. A surface of land value, a surface of road density, or a map of habitat suitability are all examples of continuous values.

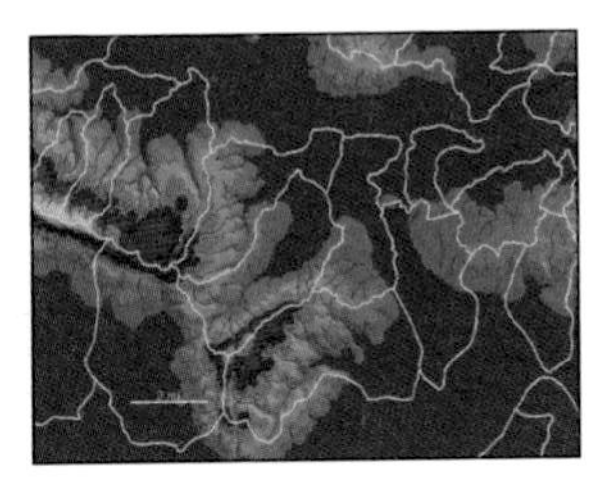

Elevation is represented as a surface of continuous values.

THE INFORMATION YOU NEED FROM THE ANALYSIS

The information you need from the analysis will also help determine which method to use.

Do you need a list, count, or summary?

You can use the GIS to find out whether an individual feature is inside an area; get a list of all the features inside an area; find out the number of features inside an area; or get a summary of what's inside an area, or each of several areas, based on a feature attribute.

For example, for parcels within a floodplain, you could:

- Find out whether a particular parcel is inside the floodplain

Land Value	Bldg Value	Bldg SqFt	Year Built	Landuse
54630	82350	1756	1952	Single Family
53720	103480	2896	1954	Single Family

- Get a list of all parcels inside the floodplain

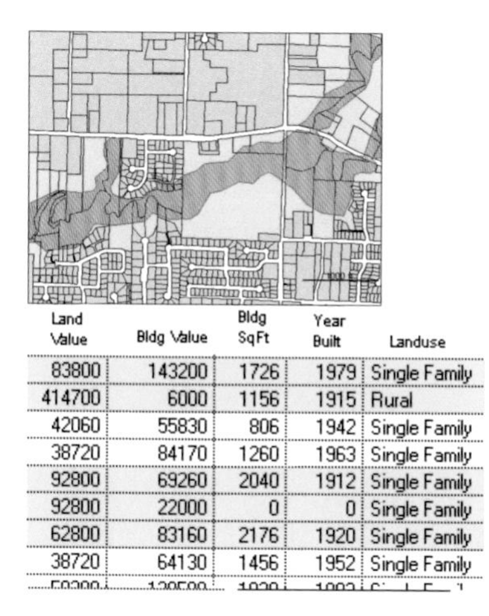

Land Value	Bldg Value	Bldg SqFt	Year Built	Landuse
83800	143200	1726	1979	Single Family
414700	6000	1156	1915	Rural
42060	55830	806	1942	Single Family
38720	84170	1260	1963	Single Family
92800	69260	2040	1912	Single Family
92800	22000	0	0	Single Family
62800	83160	2176	1920	Single Family
38720	64130	1456	1952	Single Family

- Count the number of parcels inside the floodplain

Count: 79

- Count the number of parcels of each land-use type inside the floodplain

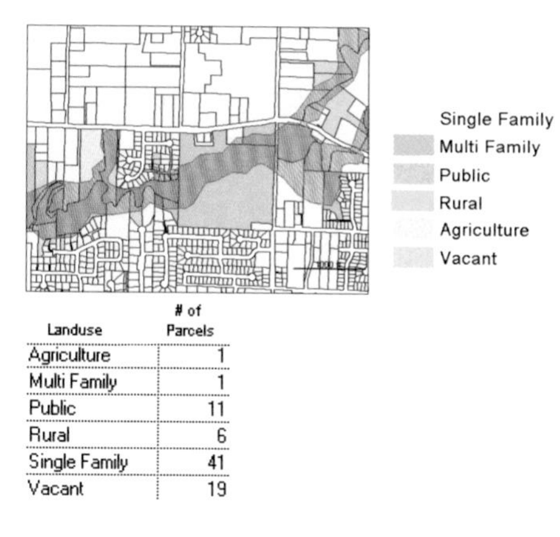

Landuse	# of Parcels
Agriculture	1
Multi Family	1
Public	11
Rural	6
Single Family	41
Vacant	19

- Sum the land value of all parcels within the floodplain

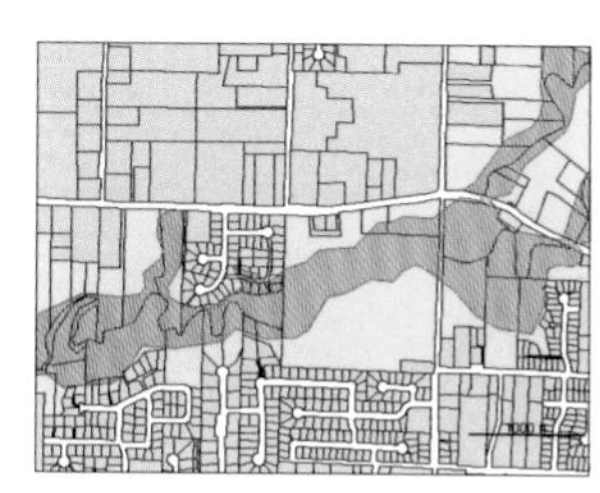

Sum: 4896330
Count: 79
Mean: 61979
Maximum: 504000
Minimum: 0

- Sum the area of each land-use type within the floodplain

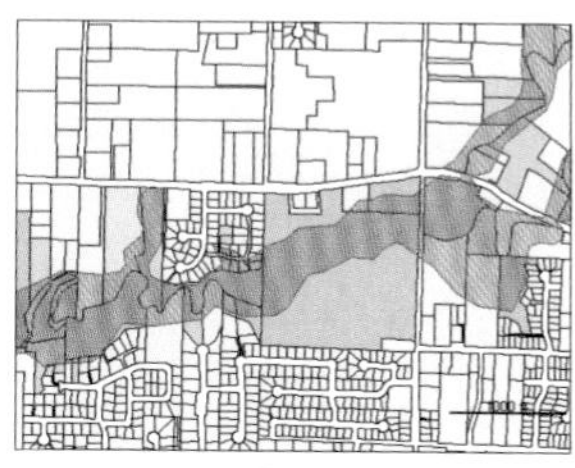

Single Family
Multi Family
Public
Rural
Agriculture
Vacant

Landuse	Total SqFt
Agriculture	261126
Multi Family	437246
Public	2592582
Rural	1055338
Single Family	2282555
Vacant	1424548

Do you need to see the features that are completely or partially inside the area?

Linear features and discrete areas might lie partially inside and outside an area. You can choose to include only features that fall completely inside, features that fall inside but extend beyond the boundary, or include only the portion of the features that falls inside the area boundary.

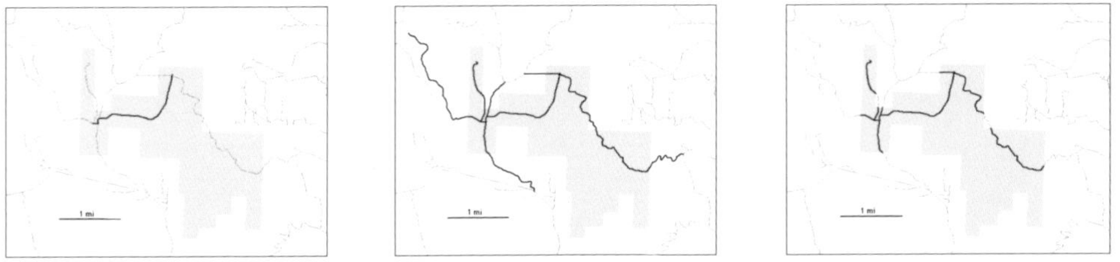

These maps show (from left to right) roads falling completely inside, roads that are partially inside, and only the portion of each road inside a protected area.

These maps show (from left to right) parcels falling completely inside a floodplain, parcels that are partially inside, and only the portion of each parcel in the floodplain.

If you need a list or count of features, you'll want to include those that are partially within the boundary. For example, when notifying surrounding property owners of a zoning change, you'd want to include parcels that are partially within the 300-foot buffer.

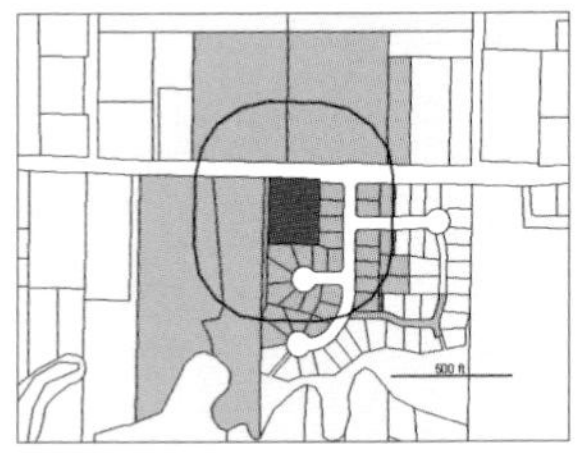

Parcels within 300 feet of a proposed zoning change

If you need to know the amount of something within the area, you'd include only the portion inside the area. For example, if you want to know the amount of each land cover type within a protected area boundary, you can use the GIS to overlay the protected area and the land cover areas. The GIS will clip out the land cover within the protected area boundary and calculate the amount of each type.

Land cover inside protected areas

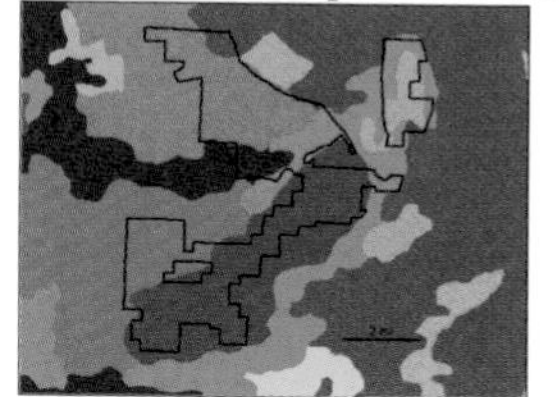

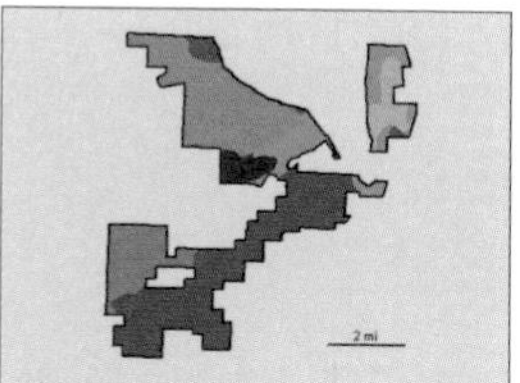

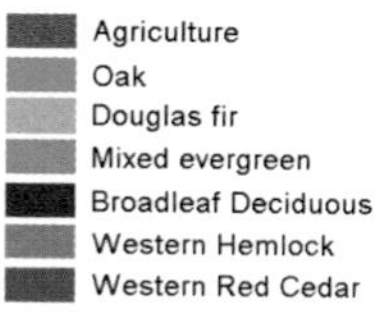

Drawing areas on top of features is a quick and easy way to see what's inside. However, you can find out what's inside in other ways that give you additional information, such as a list of the features, or summary statistics.

DRAWING THE AREAS AND FEATURES

You create a map showing the boundary of the area and the features. You can then see which features are inside and outside the area.

What it's good for

This visual approach is good for seeing whether one or a few features are inside or outside a single area.

What you need

All you need is a dataset containing the boundary of the area or areas and a dataset containing the features.

Floodplain drawn on top of parcels

SELECTING THE FEATURES INSIDE THE AREA

You specify the area and the layer containing the features, and the GIS selects a subset of the features inside the area.

What it's good for

This approach is good for getting a list or summary of features inside a single area, or a group of areas you're treating as one. It's also good for finding what's within a given distance of a feature.

What you need

You need the dataset containing the areas and a dataset with the features, including any attributes you want to summarize.

Parcels selected using a floodplain boundary

OVERLAYING THE AREAS AND FEATURES

The GIS combines the area and the features to create a new layer with the attributes of both, or compares the two layers to calculate summary statistics for each area on the fly.

What it's good for

This approach is good for finding which features are in each of several areas, or finding out how much of something is in one or more areas.

What you need

You need the data containing the areas and a dataset with the features, including any attributes you want to summarize.

Landuse	# of Parcels	Total SqFt
Agriculture	1	81046
Multi Family	1	137099
Public	11	1450742
Rural	6	420247
Single Family	43	788642
Vacant	20	814649

The floodplain has been overlaid with parcels to find out the portion of each parcel inside the floodplain.

COMPARING METHODS

Method	What it's good for	Types of features	Tradeoffs
Drawing areas and features	Finding out whether features are inside or outside an area	Locations Lines Areas Surfaces	Quick and easy, but visual only—you can't get information about the features inside
Selecting the features inside the area	Getting a list or summary of features inside an area	Locations Lines Areas	Good for getting info about what's inside a single area, but does not tell you what's in each of several areas (only all areas together)
Overlaying the areas and features	Finding out which features are inside which areas, and summarizing how many or how much by area	Locations Lines Areas Surfaces	Good for finding and displaying what's within each of several areas, but requires more processing

CHOOSING A METHOD

Follow these guidelines to choose the best method.

Draw the areas and features if you have a single area and you only need to see which features are inside.

Select the features inside the areas if you have a single area and you need a list or summary of discrete features fully or partially inside.

Overlay the areas and features if:

- You have multiple areas and you want a summary of what's inside each.
- You have a single area and you need a list or summary of discrete features, including only the portion of the features inside the area.
- You have a single area and need a summary of continuous values.

Sometimes, making a map and looking at it is all the analysis you need. By using the GIS to draw the area or areas on top of the features, you can see which discrete features are inside or outside an area, or get a sense of the range of continuous values in the area.

MAKING THE MAP

The key to this method is creating a map that makes it easy to see which features are inside the area (or areas).

Locations and lines

If you're mapping individual locations or linear features, you can draw them using a single symbol, or you can symbolize them by category or quantity. Then, draw the boundary of the area on top, usually in a thick line. If you're mapping several areas, label them so that map readers can identify each one. To further distinguish the areas, you can draw each in a different shade.

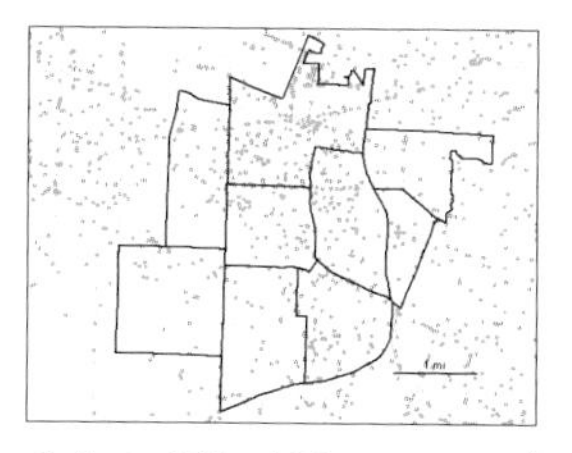

Calls to 911 within a group of neighborhoods

Calls to 911 color-coded by type of call

- Fire
- Explosion
- Medical
- Gas Leak
- Police
- Smoke
- Misc

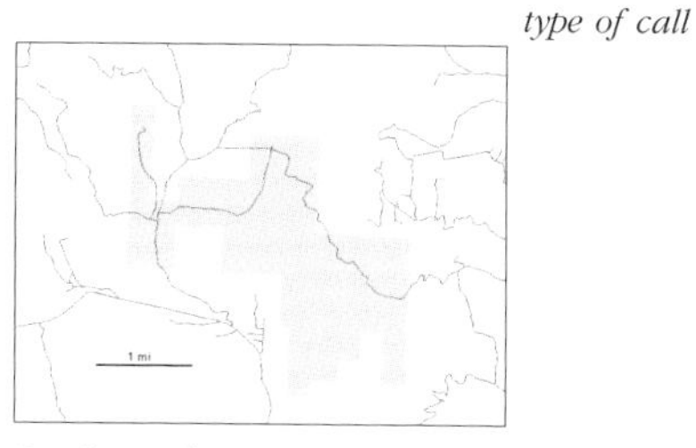

Roads inside a protected area

Discrete areas

If you want to see which discrete areas, such as parcels, are inside a single area, such as a floodplain, you have several choices. These choices depend on whether you want to emphasize the features inside or the area itself:

- Shade the outer area with a light color and draw the boundaries of the area features on top. This emphasizes which features—or portions of features—are inside.

Drawing parcel boundaries on top of the shaded floodplain emphasizes the parcels in relation to the floodplain.

- Fill the outer area with a translucent color or a pattern (for example, diagonal hatching) on top of the discrete area boundaries. This emphasizes the outer area.

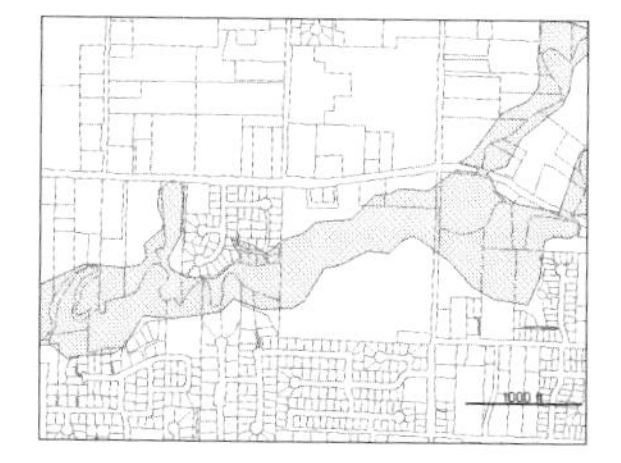

Drawing the shaded floodplain on top of the parcels emphasizes the configuration of the floodplain.

- Draw the outer area boundary with a thick line, and the discrete area boundaries with a thin line in a lighter shade or different color. Use one of these options if you're shading the discrete areas by a category or class range.

Drawing the floodplain boundary on top of parcels color-coded by land-use category shows the type of land use inside and outside the floodplain.

If you want to see which discrete area features, such as parcels, are inside several contiguous areas, such as watersheds, use contrasting shades or patterns to distinguish each area. You'll also want to label each. If you're symbolizing the discrete areas by type or quantity (for example, land parcels coded by zoning or assessed value), mapping each discrete area without its boundary will make the patterns easier to see.

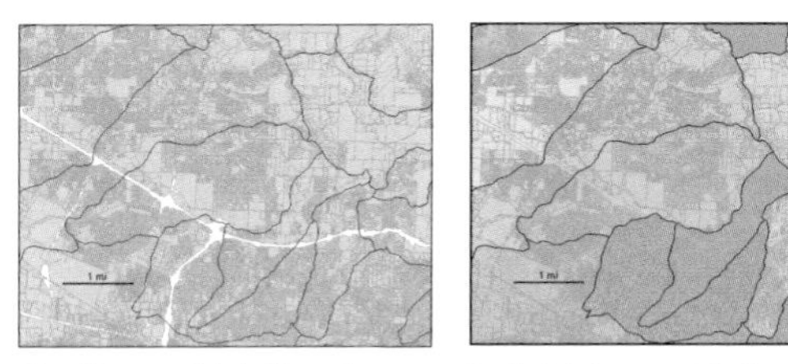

Shading and labeling the watersheds (right map) help distinguish them.

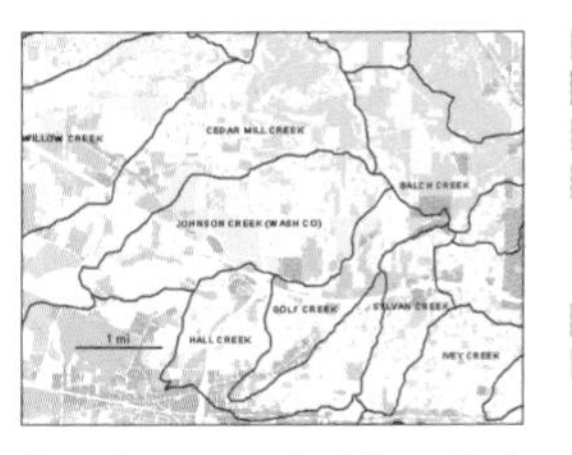

Drawing watershed boundaries with parcels color-coded by land-use category lets you see the dominant land use in each watershed.

Continuous features

If you're mapping continuous data, such as soils or elevation, draw the areas symbolized by category or quantity (as a class range), then draw the boundary of the area or areas on top. Usually, you outline the continuous data using a thin, gray line, and draw the areas on top using a thicker black line, to make the map more legible.

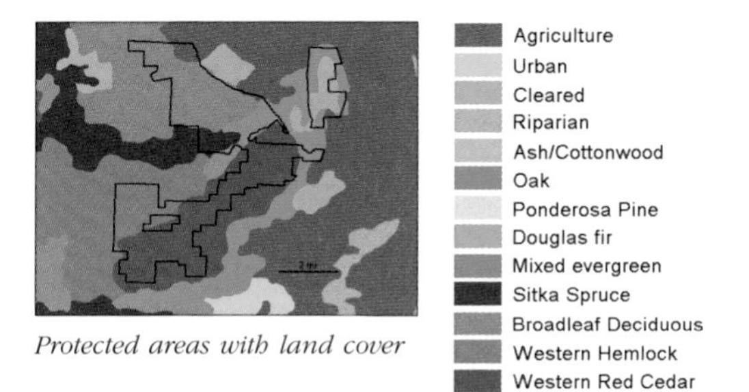

Protected areas with land cover

If you're mapping a single area, you can draw the boundary with a thick line, or shade the area with a translucent color or a pattern to highlight it. Drawing only the boundary emphasizes what's inside, while shading emphasizes the area itself. You can also place a screen over the features outside the area. This highlights the area while making it easy to see what's inside.

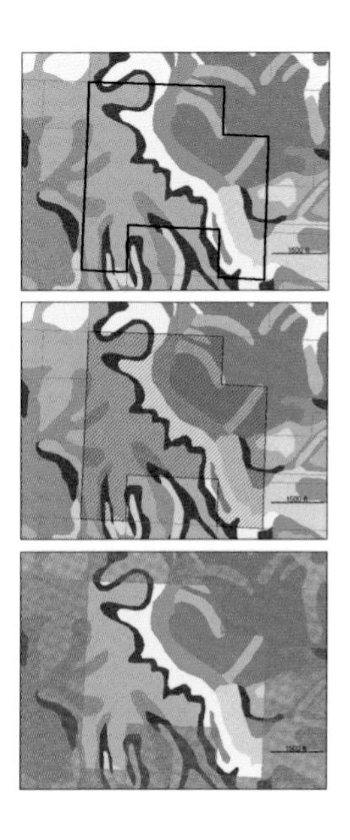

Three ways of drawing a lot boundary with soil types: outlining the lot emphasizes the soil types inside; shading the lot emphasizes the location and shape of the lot; screening the area outside the lot highlights it while showing the soil types inside.

With this method, you specify the features and the area. The GIS checks the location of each feature to see if it's inside the area, and flags the ones that are. It then highlights the selected features on the map and selects the corresponding rows in the feature set's data table. You can use the data table to get information about the features, such as a list or count. You can also summarize an attribute associated with them.

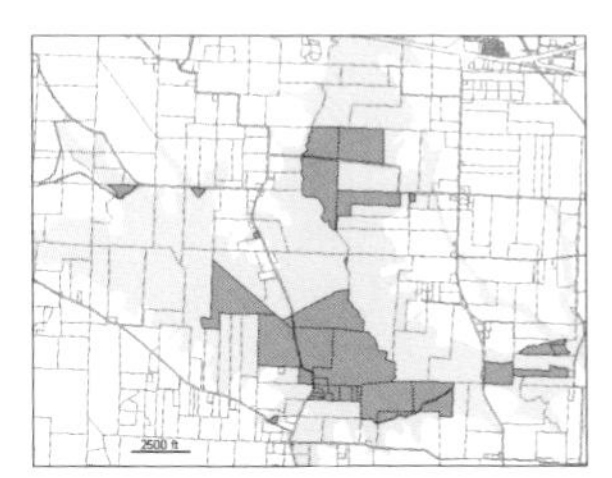

Parcels within a 100-year floodplain

Parcel ID	Land Value	Bldg Value	Acres
2N4250002100	10900	0	5.45
2N426DD01500	55250	118780	0.00
2N4320000200	0	0	26.94
2N4250002500	42200	27490	1.17
2N4320000100	0	0	14.60
2N4250002001	59850	51240	0.78
2N426DD02200	66300	213660	0.00

You can also use this method to find what's inside a set of areas you're treating as one. For example, you might want a list of calls to 911 within several adjacent neighborhoods, or the total number of eagle nests within several state parks. However, with this method, the GIS doesn't distinguish which area each feature is in—only that it's in one of them.

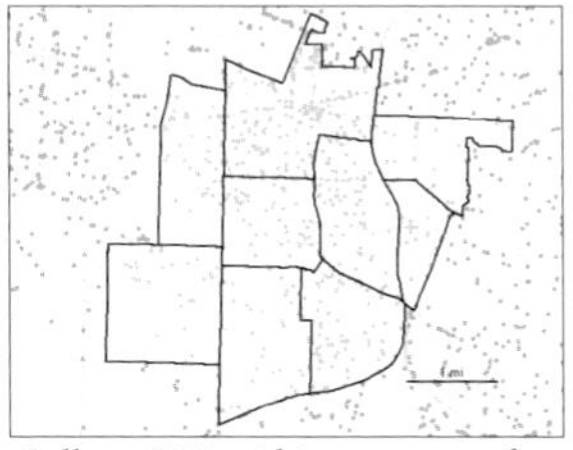

Calls to 911 within a group of neighborhoods

Call #	Date	Type
98005399	3/03/98	CFIRE
98004738	2/23/98	UND3
98010458	4/30/98	CFIRE
98012512	5/24/98	TAB1
98000759	1/09/98	SEESMOK

Geographic selection is also a quick way to find out which features are within a given distance of another feature. For example, you may need a list of residents within 500 feet of a restaurant requesting a new liquor license. You specify the location of the restaurant, the layer containing the locations of the residents, and the distance (500 feet); the GIS selects residents within that distance.

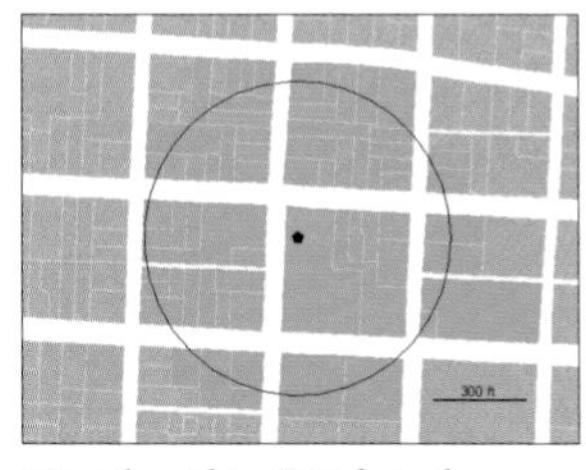

Parcels within 500 feet of a restaurant requesting a liquor license

If you have data that's already summarized by area, you can only summarize it using boundaries that fully enclose the areas. For example, if you knew the number of high school students in each census block group, you could summarize their number for each census tract, since blocks nest completely inside tracts.

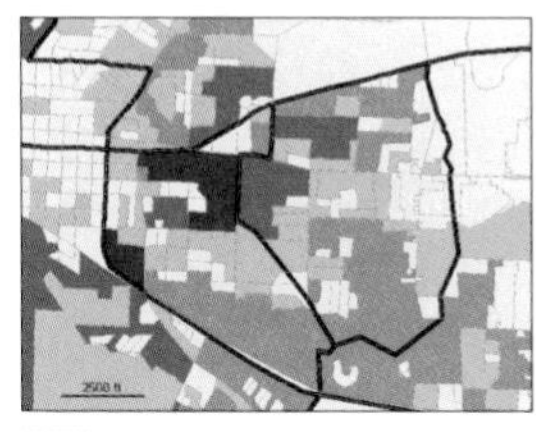

0 - 60
61 - 194
195 - 440
441 - 921
922 - 2008

Number of students by census block group

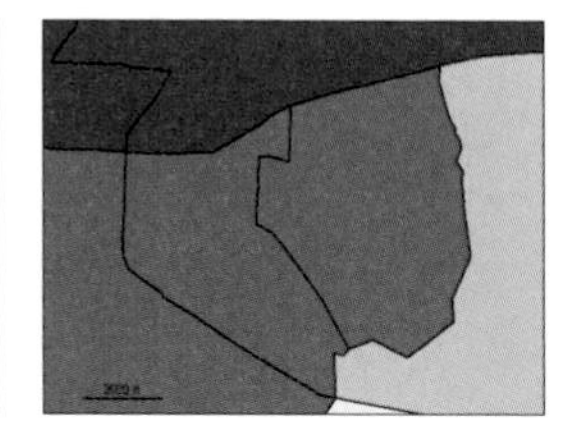

1016 - 2620
2620 - 4241
4241 - 5424
5424 - 6830
6830 - 9895

By summing the number of students in each block group, you can find the number of students in each census tract, since block groups nest within tracts.

USING THE RESULTS

You can use the GIS to create a report of the selected features. For example, you might need a list of each property within 500 feet of a proposed liquor store, so you can notify the residents.

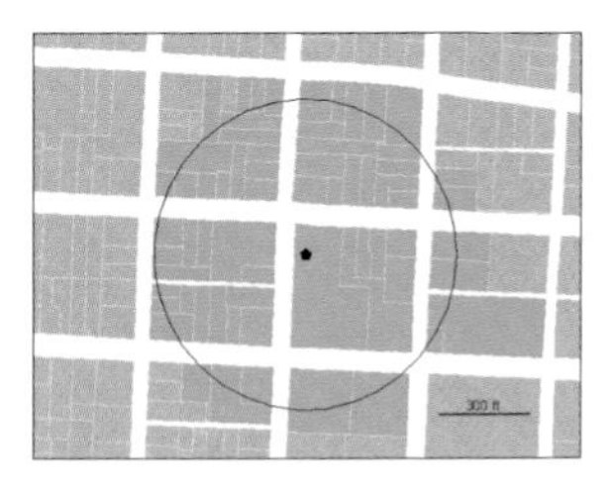

Parcels within 500 Feet of Proposed Liquor Store
3/18/99

TLID	SITEADDR	LANDUSE
1N231CD07400	554 E MAIN ST	MFR
1N231CD07300	566 E MAIN ST	MFR
1N231CD07201	574 E MAIN ST	MFR
1N231DC05000	614 E MAIN ST	SFR
1N231DC05001	622 E MAIN ST	SFR
1N231DC04900	634 E MAIN ST	SFR
1N231DC04800	650 E MAIN ST	SFR
1N231DC04700	663 E MAIN ST	MFR
1N231CD07200	105 SE 6TH AV	SFR
1N231DC05100	132 SE 6TH AV	SFR
1N231DC05200	142 SE 6TH AV	SFR
1N231CD08900	143 SE 6TH AV	SFR

You can also create statistical summaries, using the tools available in the GIS or in a spreadsheet program. Here are some of the most common summaries:

Count

A count is the total number of features inside the area, such as the number of businesses in a neighborhood.

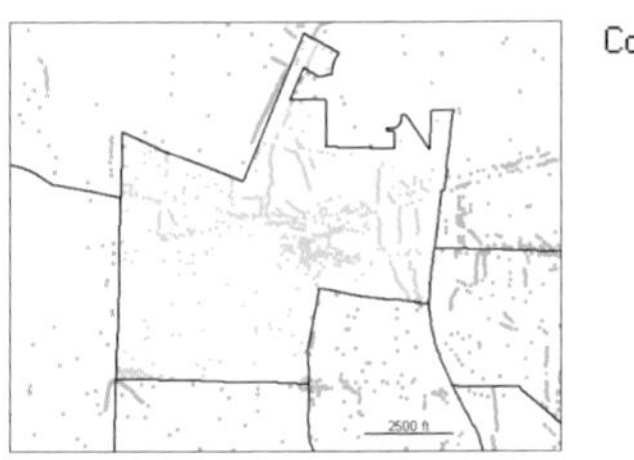

Count: 887

Total number of businesses in a neighborhood

Frequency

Frequency is the number of features with a given value, or within a range of values, inside the area, displayed as a table. An example of a frequency is the number of businesses of each type within a neighborhood. Frequency can also be displayed as a bar chart (for raw numbers) or a pie chart (for percentages).

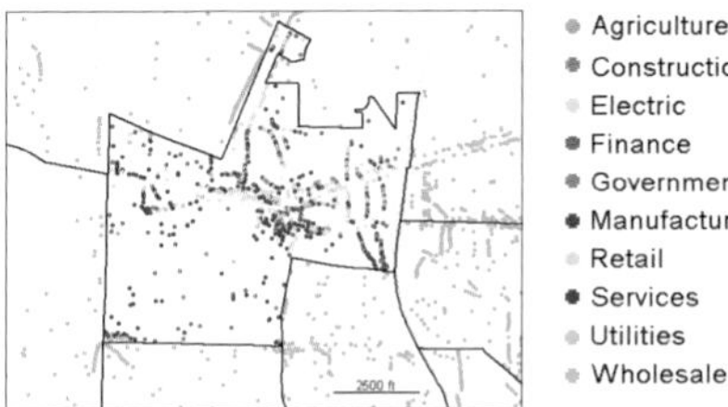

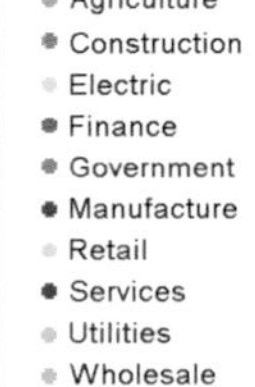

Number of businesses of each type in a neighborhood

Type	# of Businesses
Agriculture	5
Construction	30
Electric	1
Finance	136
Manufacture	30
Retail	250
Services	371
Utilities	17
Wholesale	47

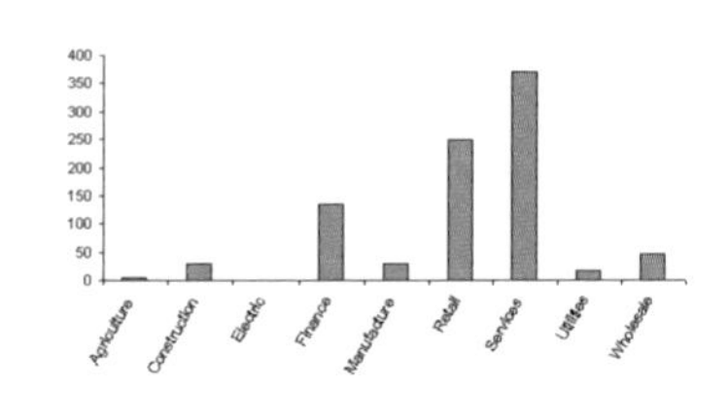

The bar chart shows the relative number of businesses of each type.

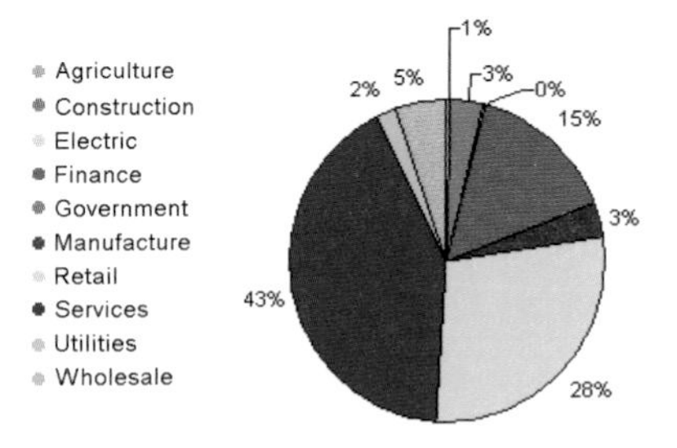

The pie chart shows what percentage of the whole each type of business composes. In this neighborhood, 43 percent of businesses are in the service industry.

A summary of a numeric attribute

The most common ones are:

- Sum. This can be the overall total, such as the total number of workers at businesses in a neighborhood, or the total by category, such as the total acreage of each land-use type within a floodplain.

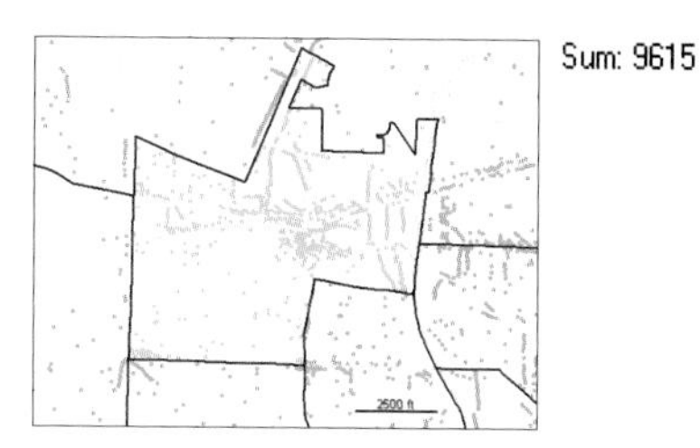

Total number of workers employed at businesses in a neighborhood

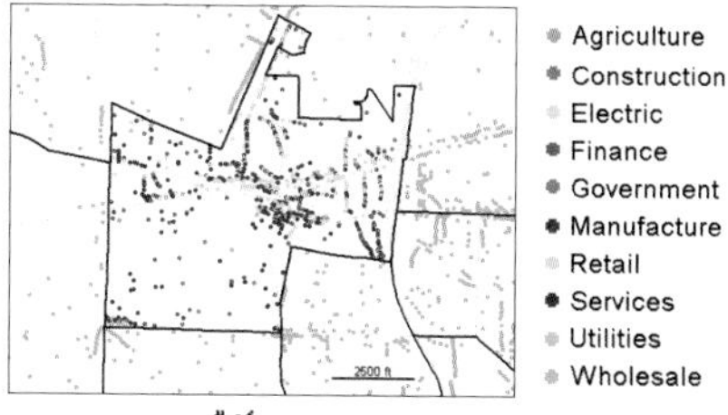

Type	# of Workers
Agriculture	49
Construction	268
Electric	72
Finance	927
Manufacture	308
Retail	4672
Services	2844
Utilities	238
Wholesale	237

Number of workers by type of business. By comparing this table with the one on the previous page, you can see that while this neighborhood has more services than retail businesses, retail employs more people.

- Average (or mean). This is the total of a numeric attribute divided by the number of features, such as the average number of workers at each business in a neighborhood. Note that very high or low numbers can skew the average.

- Median. This is the value in the middle of the range of values for an attribute (half the features have a value above this value, and half below). An example would be the value above which half the businesses in a neighborhood have more workers, and half fewer.

- Standard deviation. This is, essentially, the average amount values are from the mean. Standard deviation gives you a measure of how tightly or loosely the values are grouped.

In addition to the report or statistics, you'll also want to create a map to see which features are inside. If you want to focus only on what's inside the area, you can show only those features. However, showing all the features will provide a context for the information.

If you're drawing only the selected features, you can shade them with a single color or with a color based on an attribute value. For example, you could shade each parcel based on its land-use category. You'll also want to draw the area boundary.

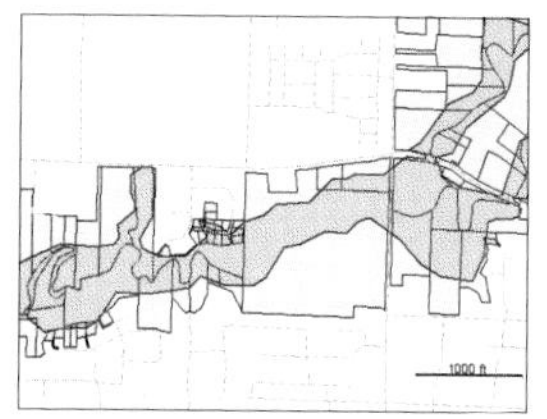

Parcels falling at least partially inside a floodplain

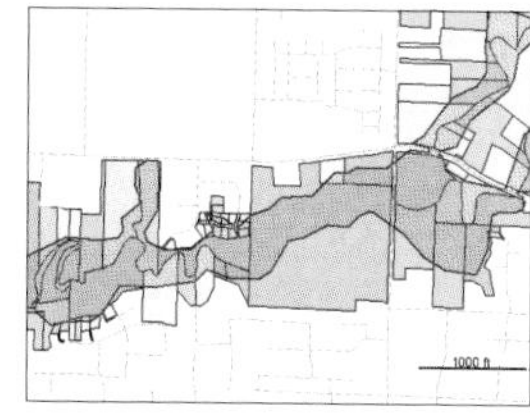

Selected parcels shaded by land-use category

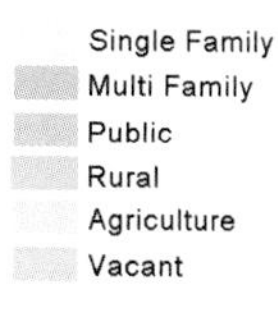

If you're drawing all the features and highlighting the ones inside, you can:

- Draw the features inside with one color and the features outside with a different, lighter color. This shows which features are inside and which are not.

Drawing all parcels and highlighting the ones inside the floodplain provides context.

- Draw the features inside based on an attribute value and draw the features outside in a single color. For discrete areas, you can simply draw the boundaries. This highlights the features inside while providing some information about what they are.

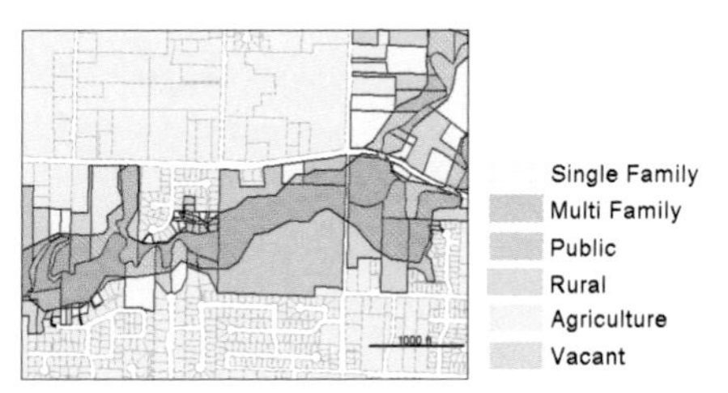

All parcels, with ones inside shaded by land use category

- Draw all the features based on an attribute value, but draw the ones outside in lighter shades. This provides the most information about the features, both inside and outside.

All parcels shaded by land-use category, with parcels inside the floodplain highlighted

This method lets you find which discrete features are inside which areas and summarize them, calculate the amount of each continuous category or class inside one or more areas, or summarize continuous values inside one or more areas.

OVERLAYING AREAS WITH DISCRETE FEATURES

The GIS tags each feature with a code for the area it falls within, and assigns the area's attributes to each feature. You can then get a list of features or a summary of an attribute value, by area. Since the attributes are permanently stored in the feature data table, you can do any number of summaries. The diagram below shows the process for calculating and mapping calls to 911 per 1,000 people, by census tract.

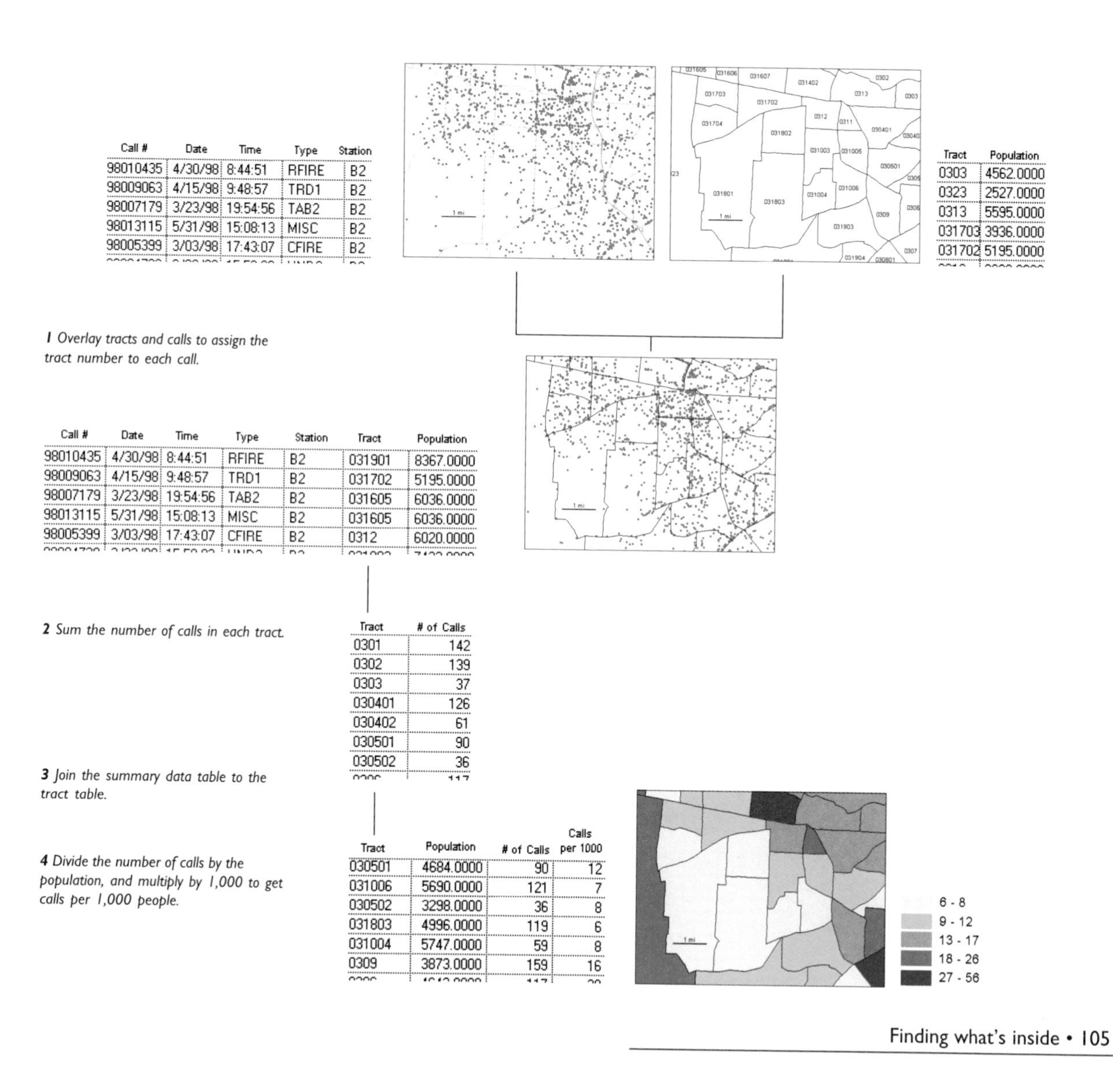

Call #	Date	Time	Type	Station
98010435	4/30/98	8:44:51	RFIRE	B2
98009063	4/15/98	9:48:57	TRD1	B2
98007179	3/23/98	19:54:56	TAB2	B2
98013115	5/31/98	15:08:13	MISC	B2
98005399	3/03/98	17:43:07	CFIRE	B2

Tract	Population
0303	4562.0000
0323	2527.0000
0313	5595.0000
031703	3936.0000
031702	5195.0000

1 Overlay tracts and calls to assign the tract number to each call.

Call #	Date	Time	Type	Station	Tract	Population
98010435	4/30/98	8:44:51	RFIRE	B2	031901	8367.0000
98009063	4/15/98	9:48:57	TRD1	B2	031702	5195.0000
98007179	3/23/98	19:54:56	TAB2	B2	031605	6036.0000
98013115	5/31/98	15:08:13	MISC	B2	031605	6036.0000
98005399	3/03/98	17:43:07	CFIRE	B2	0312	6020.0000

2 Sum the number of calls in each tract.

Tract	# of Calls
0301	142
0302	139
0303	37
030401	126
030402	61
030501	90
030502	36

3 Join the summary data table to the tract table.

4 Divide the number of calls by the population, and multiply by 1,000 to get calls per 1,000 people.

Tract	Population	# of Calls	Calls per 1000
030501	4684.0000	90	12
031006	5690.0000	121	7
030502	3298.0000	36	8
031803	4996.0000	119	6
031004	5747.0000	59	8
0309	3873.0000	159	16

What the GIS does

The GIS checks to see which area each feature is in, and assigns the area's ID and attributes to the feature's record in the data table.

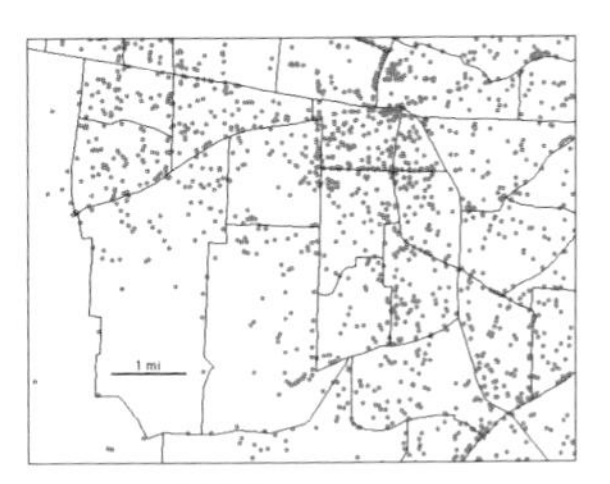

Calls to 911 by census tract

Call #	Date	Time	Type	Station	Tract	Population
98010435	4/30/98	8:44:51	RFIRE	B2	031901	8367.0000
98009063	4/15/98	9:48:57	TRD1	B2	031702	5195.0000
98007179	3/23/98	19:54:56	TAB2	B2	031605	6036.0000
98013115	5/31/98	15:08:13	MISC	B2	031605	6036.0000
98005399	3/03/98	17:43:07	CFIRE	B2	0312	6020.0000

If a line or area feature falls within two or more areas, the GIS splits the feature where it crosses the area boundary and builds new areas in a new dataset. Each new feature has the attributes of the area it falls within, in addition to its original attributes.

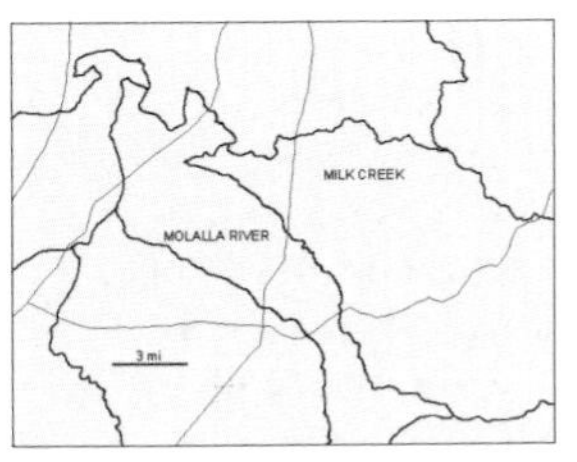

Highways in each watershed

Hwy ID	Length	Watershed
617	52140.30420	MILK CREEK
617	8766.39146	MOLALLA RIVER, LOWER
583	550.38262	BIG CREEK / GNAT CREEK
583	445.50424	BIG CREEK / GNAT CREEK

Using the results

If you're overlaying a single area, you can do the same kind of analysis you would do with geographic selection. For lines or areas, however, you're now dealing with just the portion of each feature inside the area. For example, you could calculate the total length of road only within a state park, or the percentage of each parcel within a floodplain.

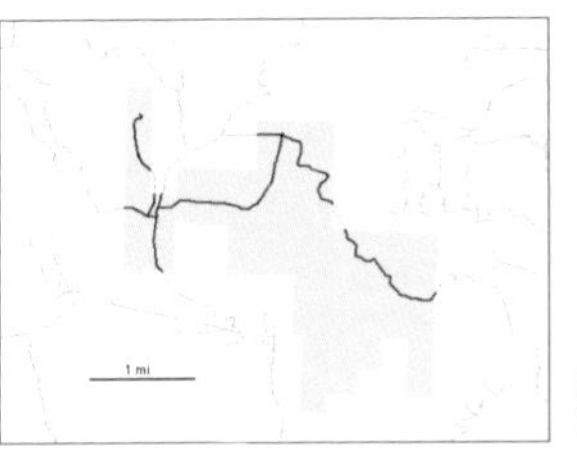

Sum: 33404.456
Count: 12

Total length of road inside a protected area

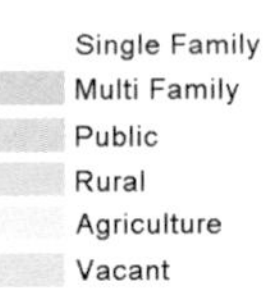

Percentage of each parcel inside a floodplain

Parcel ID	Total Area	Flood Area	% Flood Area	Landuse
1N2340004204	105250.328	33950.203	32.3	Single Family
1N2340004700	78810.703	35784.797	45.4	Vacant
1N235CB01200	104142.078	65856.688	63.2	Single Family
1N235CB01000	73228.562	52819.656	72.1	Vacant

If you're overlaying several areas on a set of features, you can summarize the features by area. If, for example, you wanted to know the number of grocery stores per capita in each census tract, you'd first summarize the feature data table to sum the number of stores in each tract. You'd then join this new summary table to the census tract data table, and calculate a new field by dividing the number of stores by the population of each tract, as shown in the following diagram.

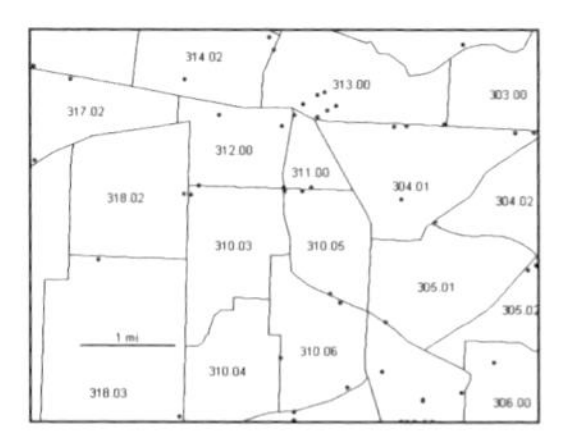

1 Overlay census tract boundaries and grocery store locations to assign a census tract number to each store.

Store	SIC	Tract	Population
Han Kuk Market	5411	312.00	6160
Daehan Oriental Foods & Gifts	5411	304.01	4013
Waremart	5411	304.01	4013
Kienows Food Stores	5411	304.01	4013

2 Sum the number of stores in each tract.

Tract	Population
303.00	4719
304.01	4013
304.02	3900
305.01	4947
305.02	3536

Tract	# of Stores
303.00	2
304.01	5
304.02	1
305.01	1
305.02	2

3 Join the summary table to the tract data table to assign to each tract the number of stores.

Tract	Population	# of Stores
301.00	8273	
314.01	10263	2
316.06	4112	1
316.07	5077	1
314.02	1052	2
303.00	4719	2
313.00	5879	8

4 For each tract, divide the number of grocery stores by the population and multiply the result by 1,000 to get the number of stores per 1,000 people.

Tract	Population	# of Stores	Stores per 1000
301.00	8273		
314.01	10263	2	0.2
316.06	4112	1	0.2
316.07	5077	1	0.2
314.02	1052	2	1.9
303.00	4719	2	0.4
313.00	5879	8	1.4

0.1 - 0.2
0.2 - 0.4
0.4 - 0.6
0.6 - 1.5
1.5 - 1.9

You can also summarize by category or value—for example, the number of businesses of each type in each tract, or the total number of employees per square mile. When summarizing numeric values by area, it's important to account for the variations in the areas (such as in size or population). To do this, you divide the value you're mapping by the size or population of each area. Chapter 3, 'Mapping the most and least,' discusses ways of comparing areas.

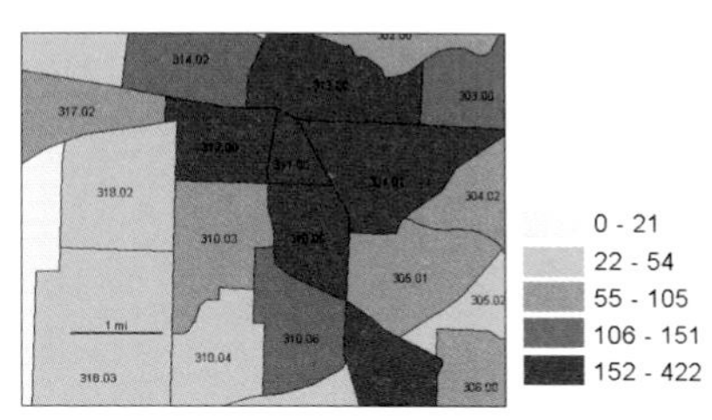

Number of workers per square mile, by census tract

Tract	Sq Mi	Workers	Workers per Sq Mile
314.02	1.6	216	137
302.00	1.8	154	86
303.00	1.2	150	120
313.00	1.4	599	422
317.03	0.9	20	21

If you're overlaying an area on data that's summarized by area (for example, overlaying a floodplain with population per census tract), you should make sure the summarized areas fall completely inside—you can't split them and be sure the values you get are accurate. Suppose that 50 percent of the land area of a census block is within a floodplain. If you multiply the population of the block by 50 percent, you can't be sure that's the number of people living in the floodplain, since most people may live in the part of the block outside the floodplain.

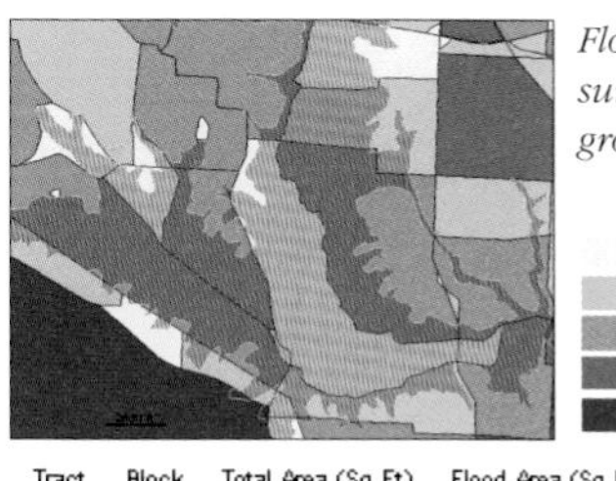

Floodplain and population summarized by census block group

Tract	Block	Total Area (Sq Ft)	Flood Area (Sq Ft)	% Flood Area	Pop
0333	239	20585456.000	12325147.5000	60	20
0327	249	3194687.250			8
0327	107	3384227.250			11

Making a map

If you have a single area, mapping individual locations is similar to mapping locations using geographic selection, described in the previous section.

If you're mapping lines or areas with a single area, you can draw just the portion of each feature inside the containing area. You'll probably want to symbolize these by category or class, and draw the portion of the features outside—along with features completely outside—using a single symbol (usually a light or neutral color).

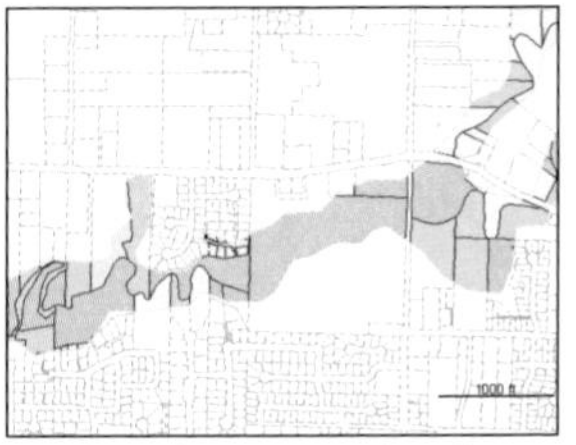

Shading the portion of each parcel inside the floodplain coded by land use highlights the land use inside the floodplain.

Shading all parcels by land use, and highlighting the portion of each parcel inside the floodplain, shows land use inside and outside the floodplain.

If you're summarizing features by area, you have a number of choices for mapping this, depending on the type of data you're summarizing. Chapter 3, 'Mapping the most and least,' describes the options.

OVERLAYING AREAS WITH CONTINUOUS CATEGORIES OR CLASSES

The GIS summarizes the amount of each category or class features falling inside one or more areas. You can get a map, table, or chart of the results.

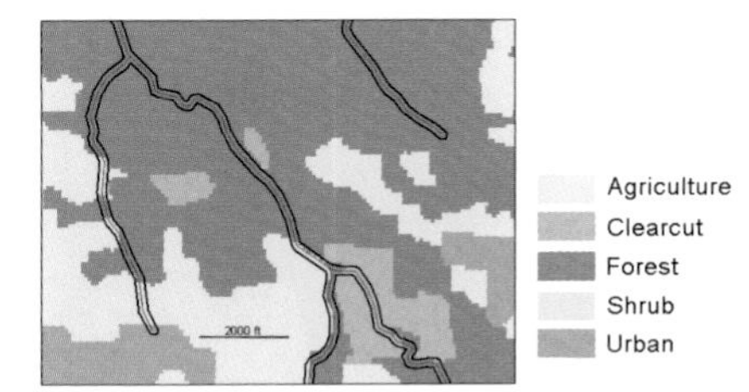

Land cover with a 200-foot stream buffer. Land cover is represented by continuous categories.

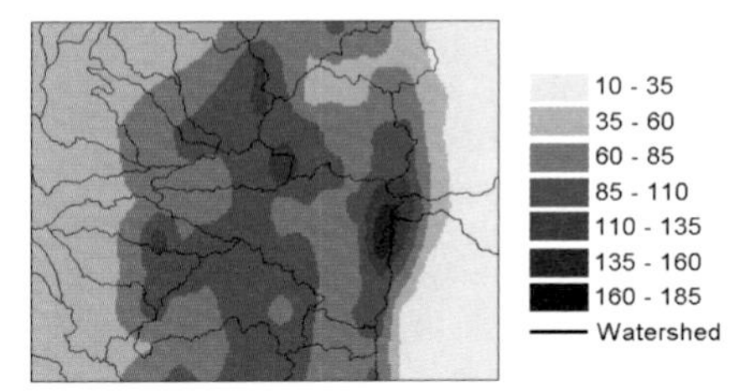

Precipitation bands with watersheds. Precipitation is represented here as classes of values.

What the GIS does

The GIS uses either a vector or raster method to overlay areas with continuous categories or classes. In some cases you'll need to choose one of these methods, while in others, the GIS just uses the best method for the data you have.

The vector method

The GIS splits category or class boundaries where they cross areas and creates a new dataset with the areas that result. Each new area has the attributes of both input layers. This is the same process as described in the previous section 'Overlaying areas with discrete features.' You then use the data table for the new layer to summarize the amount of each category in each area.

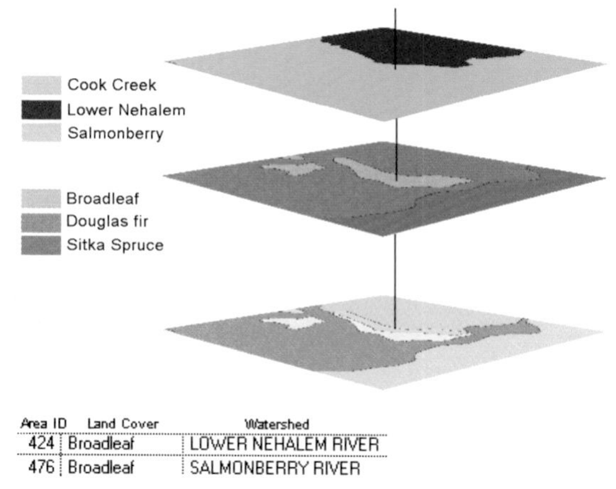

Area ID	Land Cover	Watershed
424	Broadleaf	LOWER NEHALEM RIVER
476	Broadleaf	SALMONBERRY RIVER

The area of broadleaf forest straddles a watershed boundary, so the GIS splits it into two areas—one in each watershed.

When overlaying areas on areas, you may end up with very small areas, called slivers, where borders are slightly offset. To simplify and speed up subsequent calculations, you should merge them into one of the adjacent larger areas. The GIS provides tools to automate this process.

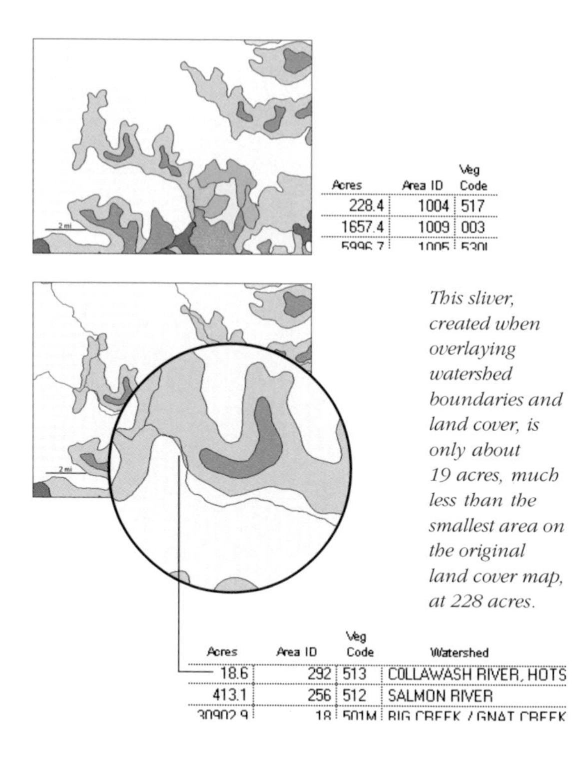

Acres	Area ID	Veg Code
228.4	1004	517
1657.4	1009	003

This sliver, created when overlaying watershed boundaries and land cover, is only about 19 acres, much less than the smallest area on the original land cover map, at 228 acres.

Acres	Area ID	Veg Code	Watershed
18.6	292	513	COLLAWASH RIVER, HOTS
413.1	256	512	SALMON RIVER

What you consider to be a sliver depends on your data. Here are some guidelines:

- Any areas with an areal extent less than the smallest area in either input dataset (sometimes referred to as the "minimum mapping unit") should be considered as a possible sliver. The minimum mapping unit is the smallest area that can be identified as a unique area on the ground. Any area smaller than this may not be a valid area.
- Consider the accuracy of your data. If you know that a boundary is accurate to within 10 feet, an area resulting from the overlay that's only 8 feet wide is probably not a valid area.
- It's a good idea to remove only very small areas at first, then manually check any remaining small areas that might be slivers. You can then have the GIS automatically remove these, or delete them manually.

The raster method

When you combine raster layers, the GIS compares each cell on the area layer to the corresponding cell on the layer containing the categories. It counts the number of cells of each category within each area, calculates the areal extent by multiplying the number of cells by the area of a cell, and presents the results in a table.

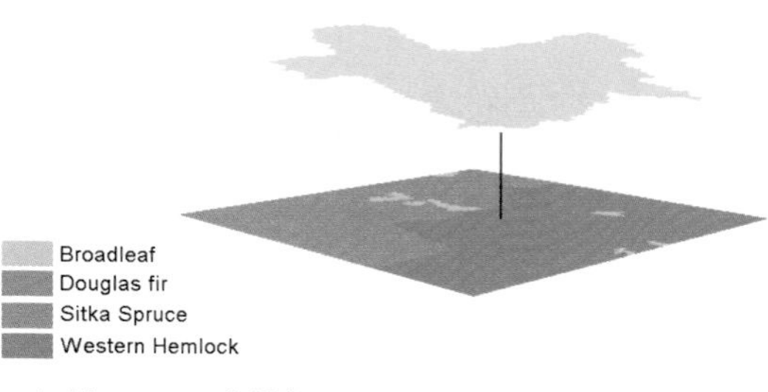

Land Cover	# of Cells	Area (Sq Ft)
Broadleaf	72	72000000
Douglas fir	11	11000000
Sitka Spruce	563	563000000
Western Hemlock	1330	1330000000

To calculate the area of each land cover type in the watershed, the GIS multiplies the number of cells of each type by the area of a cell (1,000,000 square feet, in this instance).

Should you use vector or raster overlay?

The vector method provides a more precise measure of areal extent, but requires more processing and postprocessing to remove slivers and to calculate the amount of each category in each area.

The raster method is more efficient, because it calculates the areal extent for you automatically, but can be less accurate, depending on the cell size you use. A small cell size will give more accurate results, but requires more storage space and processing power and time. Raster overlay also prevents the problem of slivers. It is often also faster, because the computation the GIS has to do is simpler.

Because raster overlay is more efficient for getting the end result—finding how much of each category or class is in each area—some GIS software makes the decision for you by converting vector layers to raster on the fly and doing a raster overlay.

Using the results

To display and analyze the results of the overlay, you'll need a table that lists the areal extent of each category within each area.

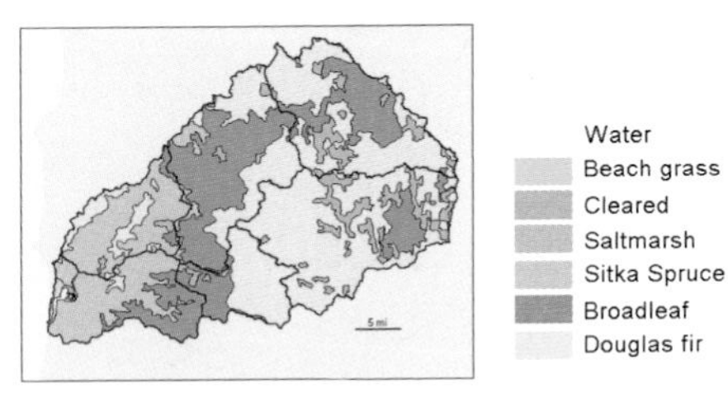

Land cover with watersheds

With raster overlay, the GIS automatically creates this table.

Watershed	Water	Beach Grass	Cleared	Saltmarsh	Sitka Spruce	Broadleaf	Douglas fir
MIDDLE NEHALEM RIVER	0.000	0.000	585236877.35	0.000	0.000	1444718477.4	2863577651.0
LOWER NEHALEM RIVER	0.000	0.000	124532963.43	0.000	304867582.62	2773070087.4	1752308092.0
NORTH FORK NEHALEM RIVER	0.000	0.000	72133847.674	0.000	1916310519.3	127935503.42	597486021.30
UPPER NEHALEM RIVER	0.000	0.000	1041177235.3	0.000	0.000	623345325.18	4556001039.4
SALMONBERRY RIVER	0.000	0.000	78258419.647	0.000	14290667.936	553933509.50	1337198213.9
COOK CREEK / LOWER NEHALEM RIVER	72133847.674	57843179.739	0.000	27220319.877	1898617311.4	930934939.79	77577911.650

Area of each land cover type in each watershed (square feet), created from a raster overlay

If you've done a vector overlay, you'll likely have several areas with the same category value inside each overlying area. To get the total amount of each category in each area, you have to summarize the category values for each area. You do this by calculating a frequency using the area identifier and category value, specifying areal extent as the field to sum. For example, you'd specify watershed name and vegetation code as the frequency items, and area as the field to sum. You'd end up with a list of each combination of watershed name and vegetation code, and the amount of land represented by that combination.

Watershed	Land Cover	Area (Sq Ft)
COOK CREEK / LOWER NEHALEM RIVER	Beach grass	58416213
COOK CREEK / LOWER NEHALEM RIVER	Saltmarsh	27965554
COOK CREEK / LOWER NEHALEM RIVER	Sitka Spruce	1907418281
COOK CREEK / LOWER NEHALEM RIVER	Broadleaf	923613244
COOK CREEK / LOWER NEHALEM RIVER	Douglas Fir	75559646
LOWER NEHALEM RIVER	Cleared	125007672
LOWER NEHALEM RIVER	Sitka Spruce	307656023
LOWER NEHALEM RIVER	Broadleaf	2767003921
LOWER NEHALEM RIVER	Douglas Fir	1757032484
MIDDLE NEHALEM RIVER	Cleared	582738056

Area of each land cover type in each watershed (square feet), created from summing the results of a vector overlay

You'd then convert this to a file that has one row for each area and a column for each category value. You can do this by modifying the table in a spreadsheet program. Or, you can select all the rows having a specific category value, create a new table, and join it to the data table for the areas. You'd then do this for each category value.

Once you have the summary table, you may also want to calculate the percentage of each category for each area, so that you can map and compare the areas based on relative amounts. This takes into account the difference in size in each area. For example, you could compare watersheds by percentage forested. To do this, add a new field to the table for each category value, and calculate values for the field by dividing the area of each category by the total areal extent of the containing area. Pie charts and stacked bar charts calculate and display the percentage for you, so you don't have to do the calculation. But having the calculated percentages will let you map each area by its percentage for a particular category.

You can display the information in the table using charts or join it to the data table for the containing areas to create maps.

Single area with multiple categories

If you're looking at how much of each category is inside a single area, you can use the table to create a bar chart showing the amount of each category in the area, or a pie chart to show what percentage of the whole each category represents.

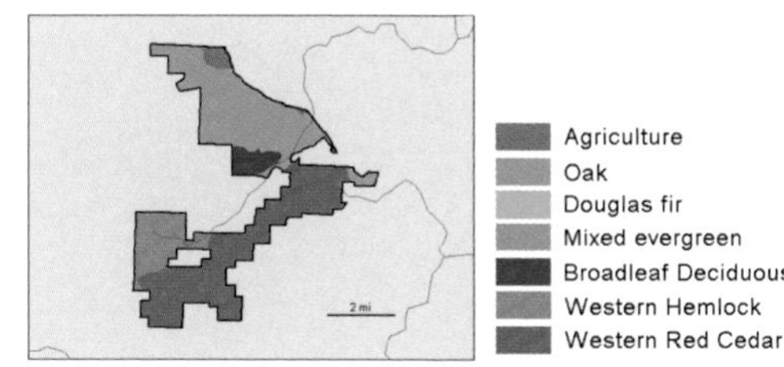

Land cover inside a single protected area

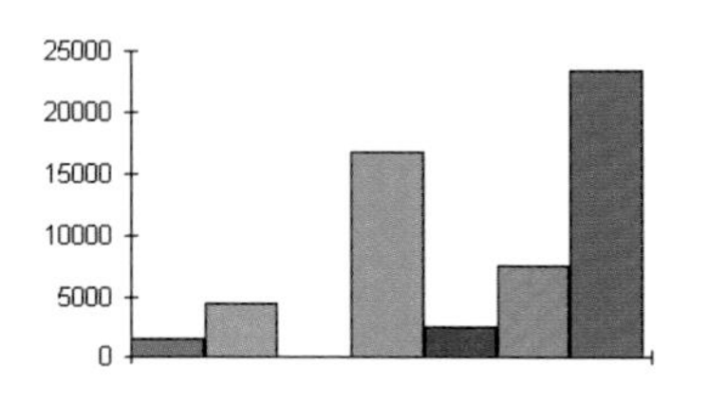

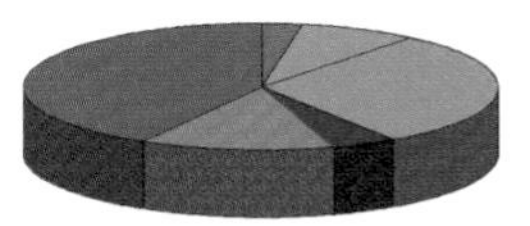

The bar chart shows the amount of each land cover, while the pie chart shows the percentage of each.

Multiple areas with a single category

If you're looking at a single category in each of several areas, a simple bar chart will show you how the areas compare. You can also map the areas based on this value. For example, you could map each watershed based on the percentage of forest in each.

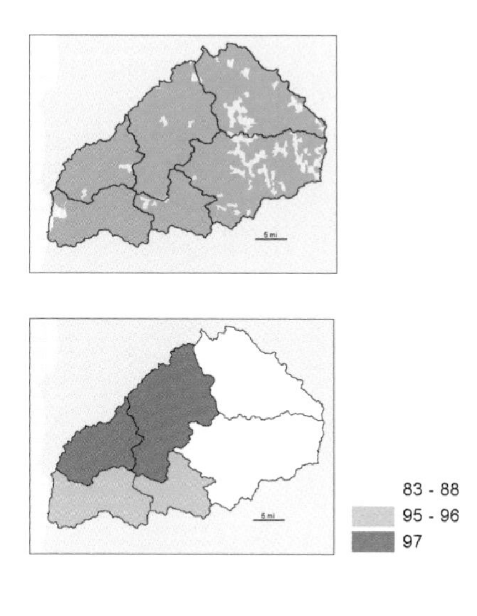

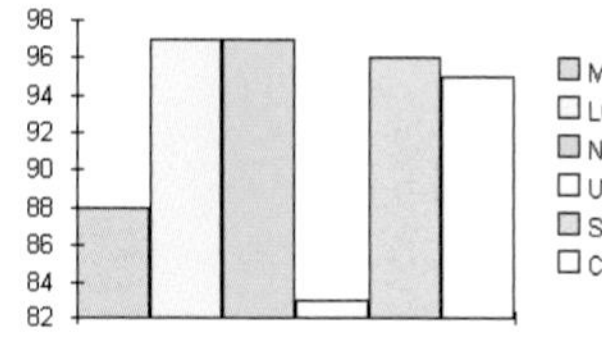

The bar chart compares the percentage of forest in each watershed.

Multiple areas with multiple categories

If you're looking at each category in each area, you can create a histogram showing multiple side-by-side bars. The chart graphically shows you the makeup of each area—that is, the amount of each category in the area—as well as how areas compare for each category. This works best for just a few classes and areas. Otherwise, it becomes difficult for readers to make the comparisons. Another way to do this is to create a clustered or stacked bar chart.

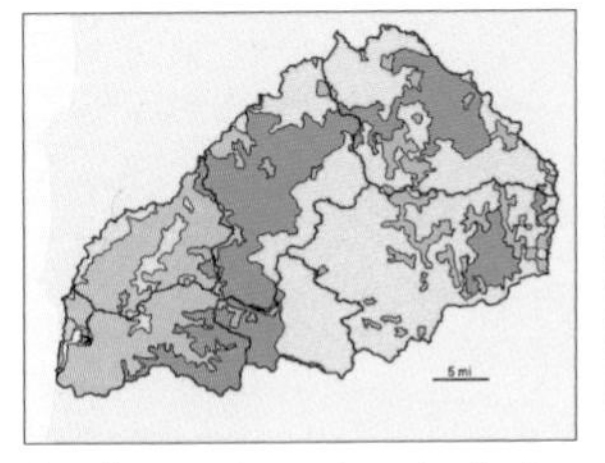

Water
Beach grass
Cleared
Saltmarsh
Sitka Spruce
Broadleaf
Douglas fir

Land cover in each watershed

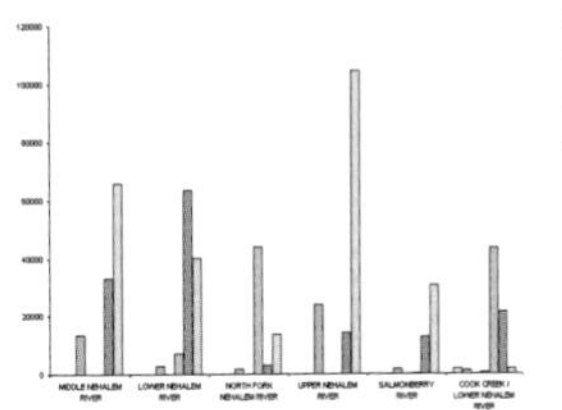

The clustered bar chart shows the amount of each land cover type in each watershed, so you can compare a specific type.

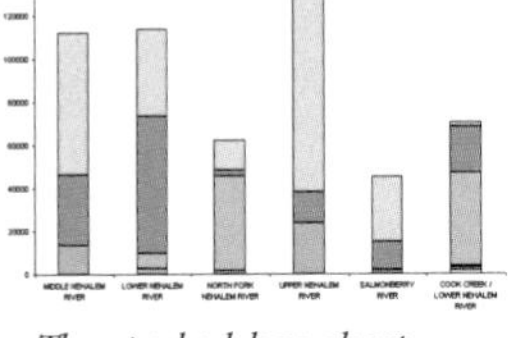

The stacked bar chart shows the relative land area of each watershed, and the land cover composition of each.

You can also place a pie or bar chart in each area, although this also works best for just a few areas and a few categories. You should also make sure the charts are large enough to read clearly.

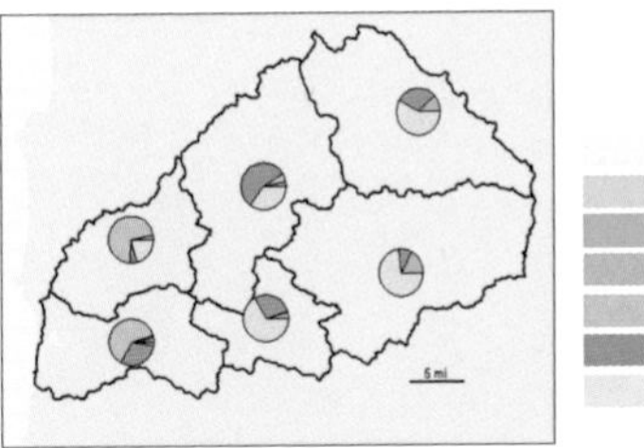

Water
Beach grass
Cleared
Saltmarsh
Sitka Spruce
Broadleaf
Douglas fir

The pie charts show the percentage of each land cover type in each watershed.

You can use this same method to visualize relationships—that is, whether things are likely to be found together more than chance would suggest—without getting into statistical analysis like correlation. For example, instead of summarizing vegetation type by unique areas, such as watersheds, you could summarize by precipitation classes to find the mean precipitation for each vegetation type, and see if there's a relationship between amount of precipitation and type of vegetation in a place. However, you'd need to do additional statistical analysis to establish that there actually is a cause-and-effect relationship.

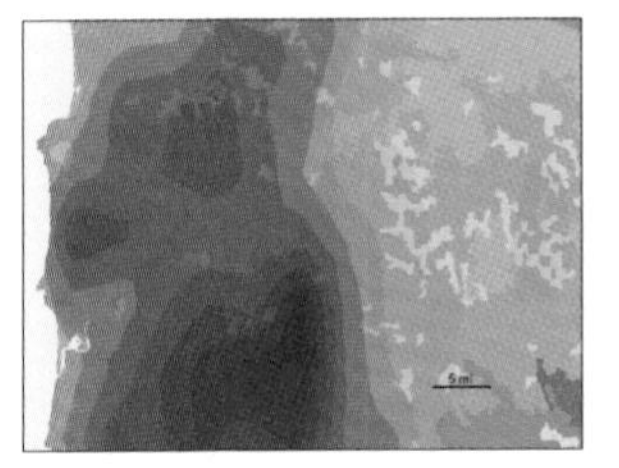

Precipitation (inches per year) and land cover

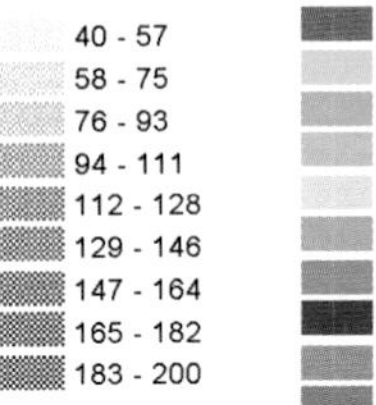

Agriculture
Urban
Beach grass
Cleared
Saltmarsh
Ash/Cottonwood
Spruce/Hemlock
Oak woodland
Broadleaf
Douglas fir

Land Cover	Acres	Min	Max	Range	Mean	Std Dev
Agriculture	41946	0	90	90	90	8
Urban	4796	0	43	43	43	17
Beach grass	3640	75	98	23	23	6
Cleared	86516	46	168	122	122	28
Saltmarsh	2578	83	102	19	19	5
Ash/Cottonwood	3156	50	70	20	20	5
Spruce/Hemlock	249738	75	177	102	102	17
Oak woodland	11482	0	51	51	51	12
Broadleaf	328006	50	200	150	150	49
Douglas fir	518630	0	200	200	200	33

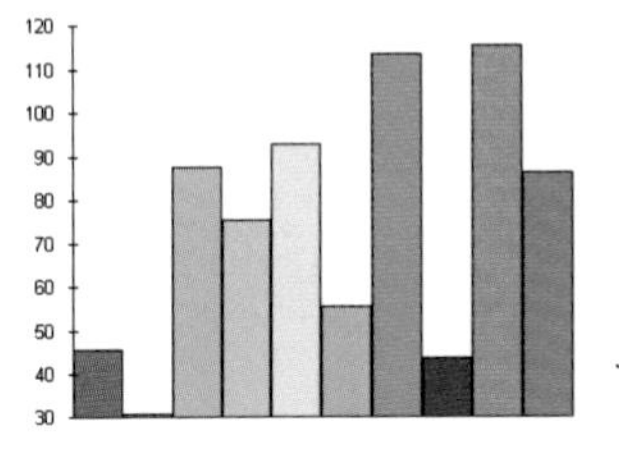

Mean precipitation for each land cover type. Agriculture and oak woodland tend to be found in areas with less rainfall, while spruce/hemlock and broadleaf forests are found in areas with more rainfall.

OVERLAYING AREAS WITH CONTINUOUS VALUES

If you have a layer of continuous values, such as elevation, you can have the GIS summarize the values and create a map or table of summary statistics for each area. These include the mean, minimum value, maximum value, value range (the difference between the minimum and maximum), standard deviation, and sum.

For example, if you overlay an elevation surface and a watershed layer, the GIS calculates the mean elevation within each watershed. It also calculates the minimum elevation, maximum elevation, elevation range, and standard deviation around the mean. You could then compare watersheds based on these values, or select watersheds that meet a certain criterion.

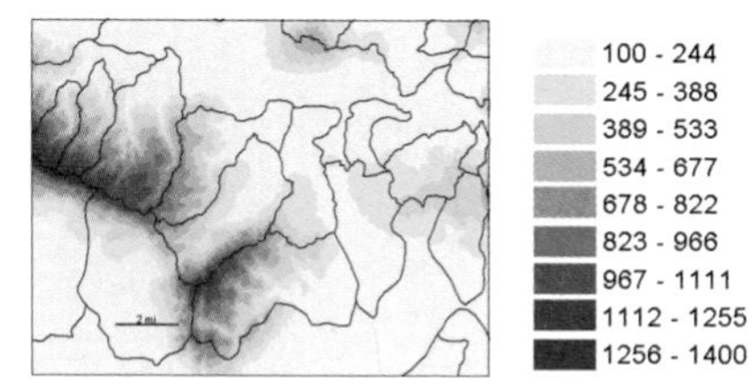

Elevation (in feet) and watershed boundaries

Watershed	Min	Max	Range	Mean
SPRING BROOK CREEK	100.0000	1200.0000	1100	328
AYERS CREEK	200.0000	1400.0000	1200	843
MCFEE CREEK	200.0000	1359.6935	1160	452
HEATON CREEK	200.0000	1300.0000	1100	601
JAQUITH CREEK	200.0000	1400.0000	1200	687
PECAN CREEK	118.7155	700.0000	581	386

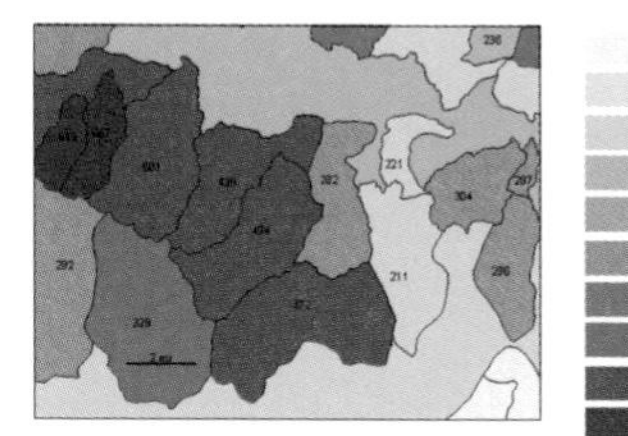

Watersheds classified by mean elevation

What the GIS does

The GIS first finds out which cells fall within each area. Once it's identified these, it calculates the statistic for the characteristic you're interested in and assigns the value to each cell it's identified. It then continues to the next area and repeats the process.

Using the results

You can create a chart from the table to compare areas based on a particular statistic.

You can also join the summary table to the data table for the areas, and map the areas by any of the summary statistics. Some GIS software, such as ArcInfo, lets you automatically create a chart or map at the same time as it creates the table.

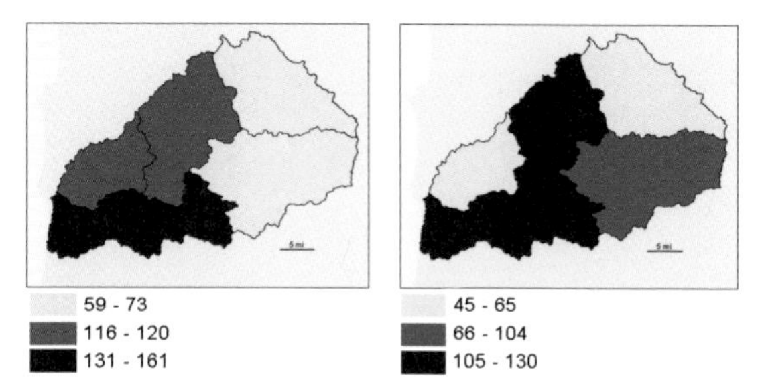

Watersheds by mean precipitation in inches per year (left), and by precipitation range (right). A wider range indicates a greater variation in rainfall across the watershed.

Watershed	Min	Max	Range	Mean
MIDDLE NEHALEM RIVER	51	96	45	59
LOWER NEHALEM RIVER	60	190	130	116
NORTH FORK NEHALEM RIVER	97	162	65	120
UPPER NEHALEM RIVER	49	153	104	73
SALMONBERRY RIVER	95	200	105	161
COOK CREEK / LOWER NEHALEM	85	190	105	131

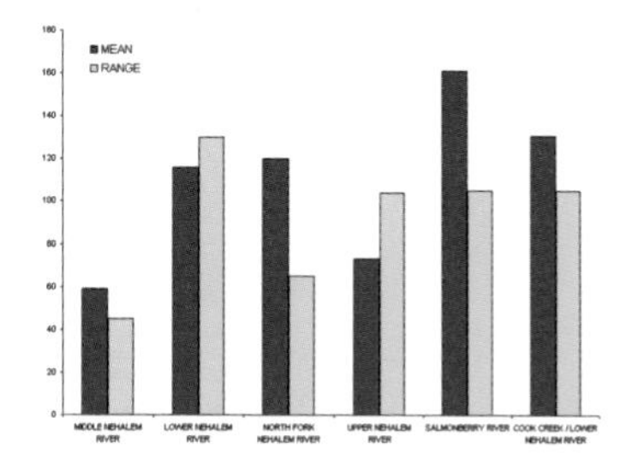

The bar chart compares mean precipitation to precipitation range. Watersheds with a high mean and a low range have fairly consistent rainfall across the watershed.

Finding what's nearby

Finding what's nearby lets you see what's within a set distance or travel range of a feature. This lets you monitor events in an area, or find the area served by a facility or the features affected by an activity.

In this chapter:

Using GIS, you can find out what's occurring within a set distance of a feature. You can also find out what's within traveling range.

Finding what's within a set distance identifies the area—and the features inside that area—affected by an event or activity. For example, a city planner may need to notify all residents within 500 feet of a proposed liquor store. Finding what's within a set distance also lets you monitor activity in the area. For example, a state forester would monitor logging to make sure it doesn't occur within a 100-meter buffer along streams.

Traveling range is measured using distance, time, or cost. Finding what's within traveling range of a feature can help define the area served by a facility. A fire chief would want to know which streets are within a three-minute drive of a fire station, or a retail analyst might want to find out how many people live within a 15-minute drive of the proposed site for a new store.

Knowing what's within traveling range can also help delineate areas that are suitable for, or capable of supporting, a specific use. For example, a wildlife biologist might map the area within a half-mile of streams, and combine this with vegetation type, slope, and other factors to identify prime deer habitat.

MAP GALLERY

The Department of Fire and Rescue in Prince William County, Virginia, created a map showing areas within 5, 10, and 15 minutes of fire stations. Policy makers used the information to help them decide where to build new stations.

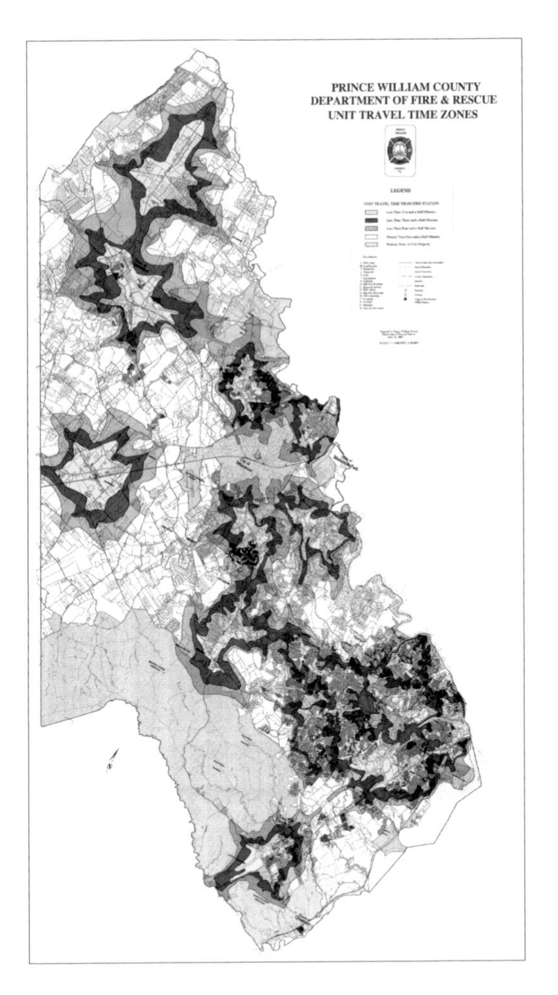

Analysts at the City of Seattle mapped one-mile buffers around branch library locations to show the land use surrounding each and highlight where new branches might be needed. The map was used during city council discussions of a proposed bond issue to fund new libraries.

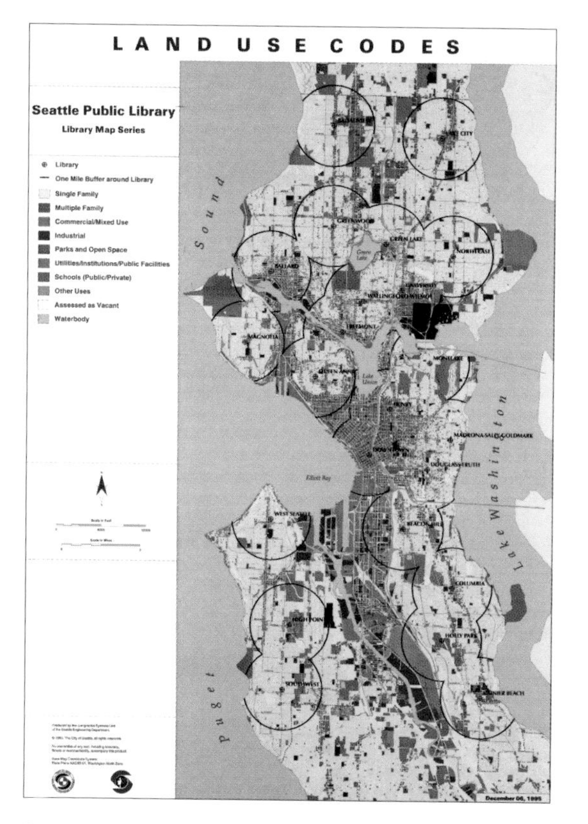

Decision Support Services, a consulting firm in Brooklyn, New York, created a map that shows the locations of customers color-coded by which software store they are nearest to, based on drive time. The information was used to support a direct mail campaign for a software company.

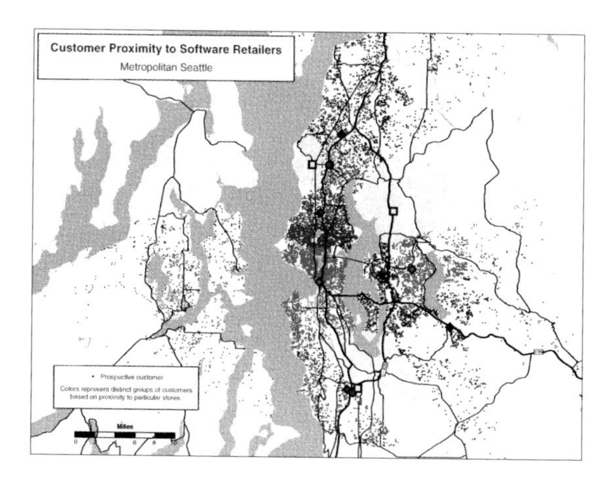

Analysts at the City of Winston–Salem, North Carolina, created a spider diagram to show which fire company first responded to each fire, over the period from 1990 to 1998. The map shows how far and in which direction each company traveled to fires, and where companies responded to fires outside their home territory. The map was used to determine which fire companies would provide backup for each part of the city.

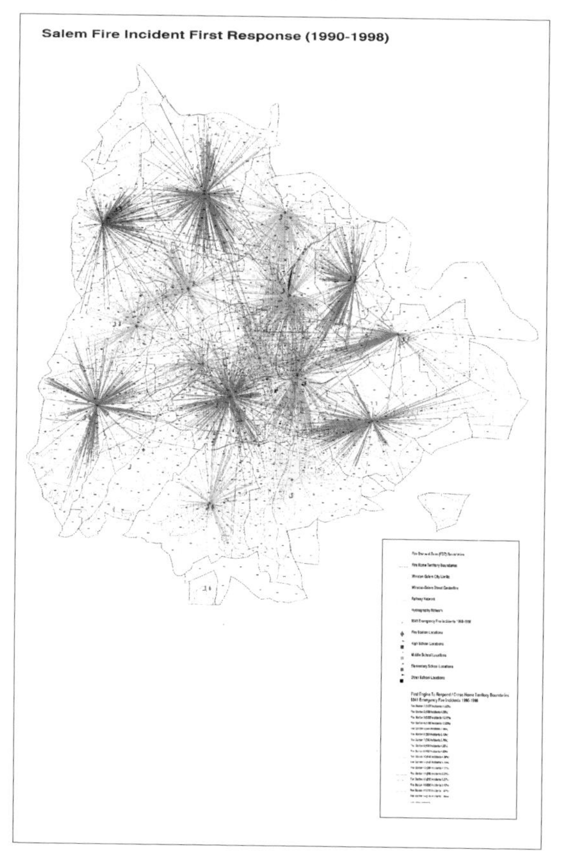

To find what's nearby, you can measure straight-line distance, measure distance or cost over a network, or measure cost over a surface. Deciding how to measure "nearness" and what information you need from the analysis will help you decide which method to use.

DEFINING AND MEASURING NEAR

What's nearby can be based on a set distance you specify, or on travel to or from a feature. If travel is involved, you can measure nearness using distance or travel cost.

Is what's nearby defined by a set distance, or by travel to or from a feature?

The surrounding features may simply be within a source feature's area of influence. This includes, for example, properties within 300 feet of a proposed zoning change, or forest within a 100-foot stream buffer. In these cases, there is no movement between the source and surrounding features. An area of influence is usually measured using straight-line distance.

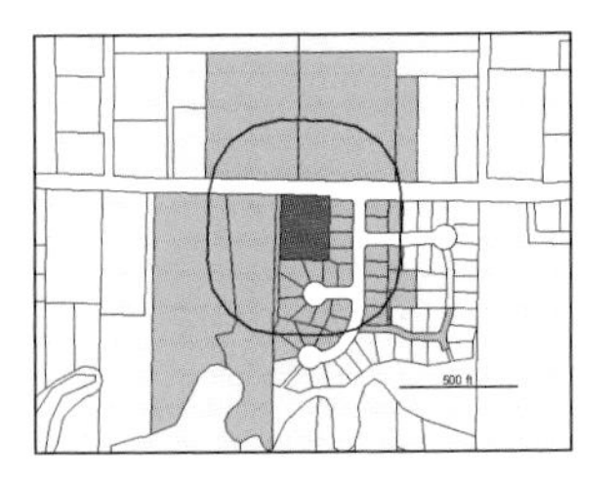

Agriculture
Clearcut
Forest
Closed shrub
Open shrub
Scattered shrub
Urban

Alternatively, there may be movement or travel between the source and the surrounding features. For example, people driving to a store, or a fire truck traveling from the station to a fire. Travel can be measured over a geometric network, such as streets or transmission lines, or over land, such as deer walking to a stream.

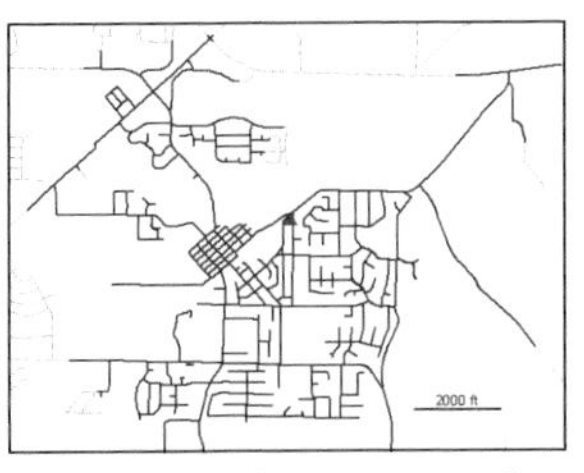

Streets within three minutes of a fire station

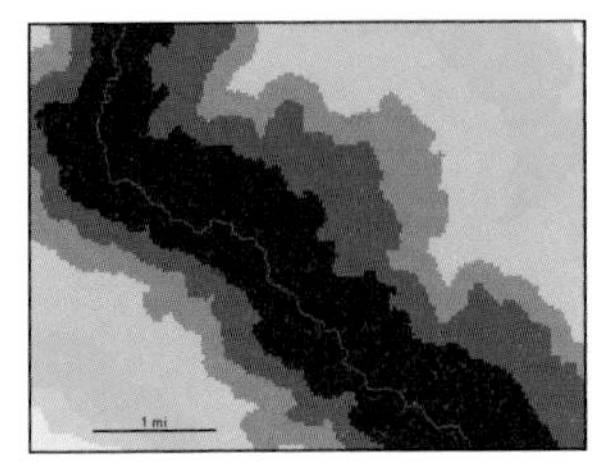

Cost of travel over various land cover types

Are you measuring what's nearby using distance or cost?

Distance is one way of defining and measuring how close something is. But nearness doesn't have to be measured using distance. You can also measure what's nearby using cost. Time is one of the most common costs—it takes longer, for example, for customers to get to a store through heavy traffic. Other costs include money (such as the operating cost per mile for a delivery van), and effort expended (for instance, for a deer walking through thick underbrush versus open forest to reach a stream). These are often referred to as "travel costs."

If you're mapping what's nearby based on travel, you can use distance or cost. Mapping travel costs gives you a more precise measure of what's nearby than mapping distance, but requires more data preparation and processing.

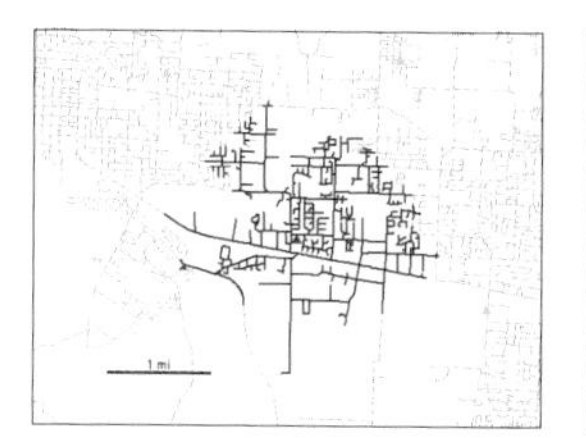

Streets within ¾ of a mile of a fire station

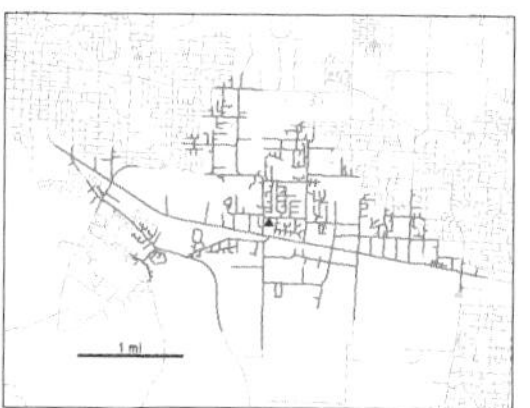

Streets within three minutes of a fire station

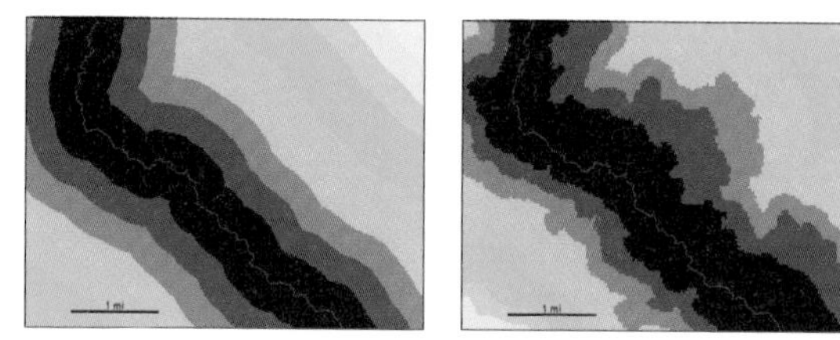

Distance from a stream versus travel cost based on land cover

For example, a retail analyst interested in finding out how many customers are within a 20-minute drive of stores could draw a circle with a radius of 5 miles around each store, calculate the area within 5 miles of each store along streets, or calculate the area within 20 minutes along streets. Using the GIS to draw a 5-mile radius around each store is easier and faster than setting up a street network with travel time along streets, but would give a less accurate count of the number of customers. A fire chief, on the other hand, would want to know as precisely as possible which streets are within three minutes of each fire station, and would use the GIS to measure actual travel time.

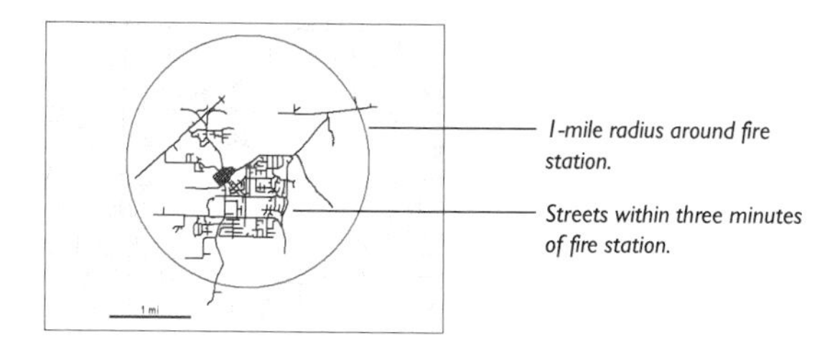

THE INFORMATION YOU NEED FROM THE ANALYSIS

Knowing the information you need will help you choose the best method for your analysis.

Do you need a list, count, or summary?

Once you've identified which features are near a source, you can get a list of the features, a count, or a summary statistic based on a feature attribute.

A list. An example of a list is the parcel ID and address of each lot within 300 feet of a road repair project.

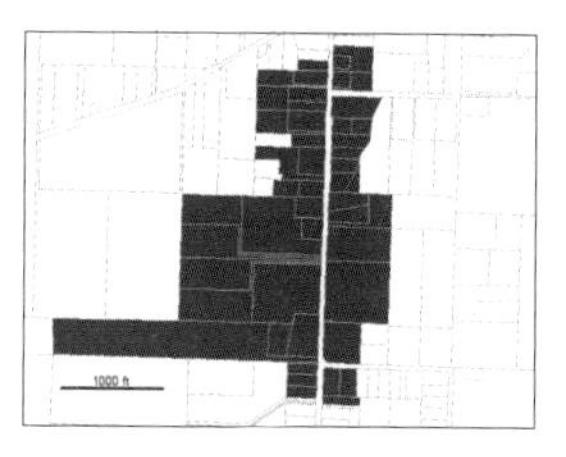

Parcel ID	Address
22E26D 00101	16471 S BRADLEY RD
22E26D 00103	16491 S BRADLEY RD
22E26D 00102	16511 S BRADLEY RD
22E26D 00104	16461 S BRADLEY RD
22E26D 00202	16551 S BRADLEY RD

A count. The count can be a total, or a count by category. For example, you could get the total number of calls to 911 within a mile of a fire station over a six-month period, or the number of calls by type.

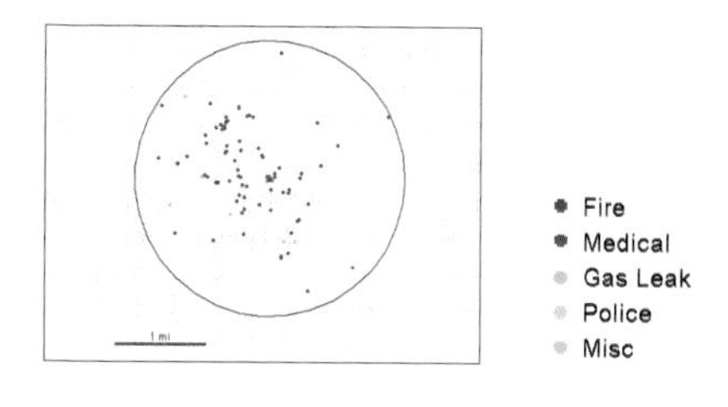

Call Type	# of Calls
FIRE	14
MEDICAL	110
GAS LEAK	2
POLICE	1
MISC	3

A summary statistic. This can be either:

- A total amount—for example, the number of acres of land within a stream buffer
- An amount by category—for example, the number of acres of each land cover type (forest, meadow, and so on) within a stream buffer

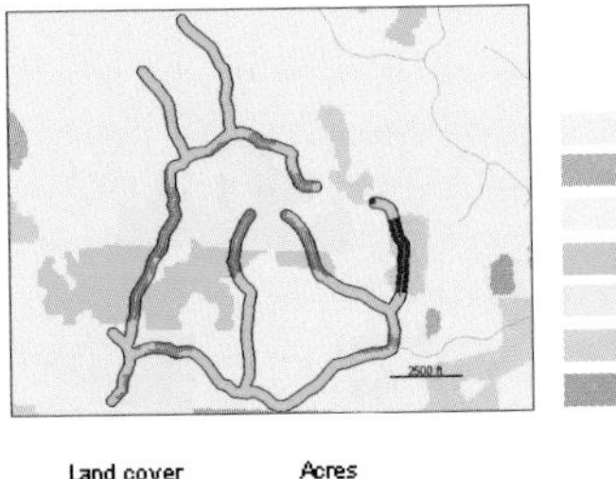

Land cover	Acres
Clearcut	4.6
Scattered shrub	23.8
Closed shrub	35.8
Open shrub	84.1
Forest	302.2

- A statistical summary, such as an average, minimum, maximum, or standard deviation—for example, the mean square footage of buildings within three minutes of each fire station

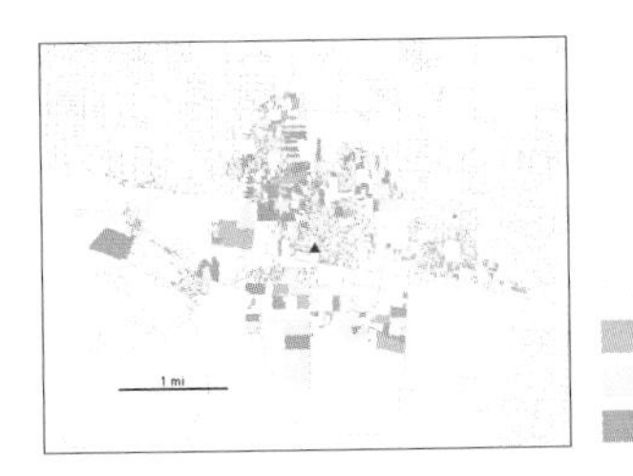

Mean: 1318
Maximum: 19992

How many distance or cost ranges do you need?

You can specify a single range or several ranges. For example, you may want to know how many customers are within 3 miles of your store, or within 1, 2, and 3 miles.

If you're specifying more than one range you can create either inclusive rings or distinct bands.

Inclusive rings

Inclusive rings are useful for finding out how the total amount increases as the distance increases. For example, you can calculate the total number of customers within 1,000 feet, the total within 2,000 feet, and the total within 3,000 feet of a store, and see how the number increases.

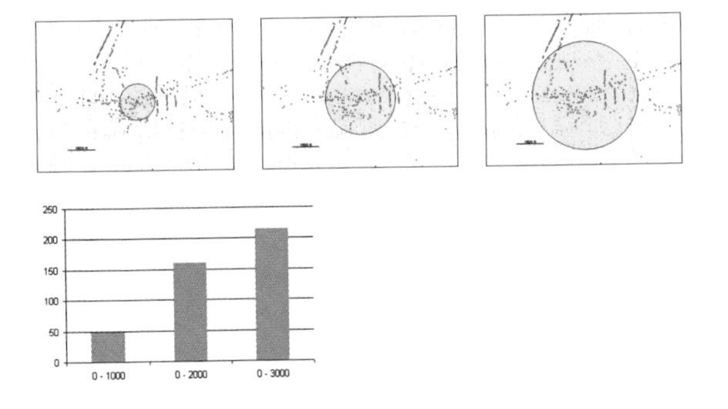

Distinct bands

Bands are useful if you want to compare distance to other characteristics. For example, you could find the number of customers within 1,000 feet, the number between 1,000 and 2,000 feet away, and the number between 2,000 and 3,000 feet away from a store. You could then see if people living in the 0- to 1,000-foot band spend more per person than people living in the 2,000- to 3,000-foot band, and so on.

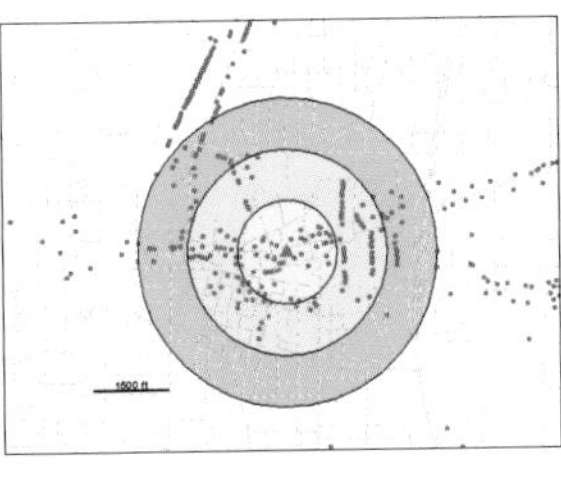

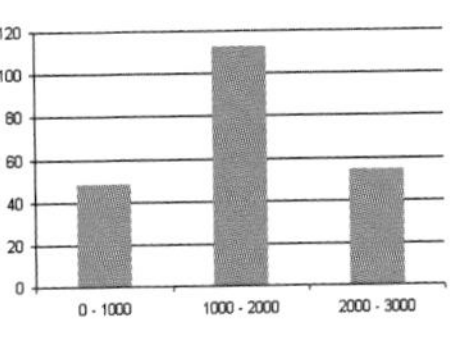

A quick way of finding what's nearby is to use straight-line distance. However, measuring distance or cost over a network, or cost over a surface, can give you a more accurate measure of what's nearby.

STRAIGHT-LINE DISTANCE

With straight-line distance, you specify the source feature and the distance, and the GIS finds the area or the surrounding features within the distance.

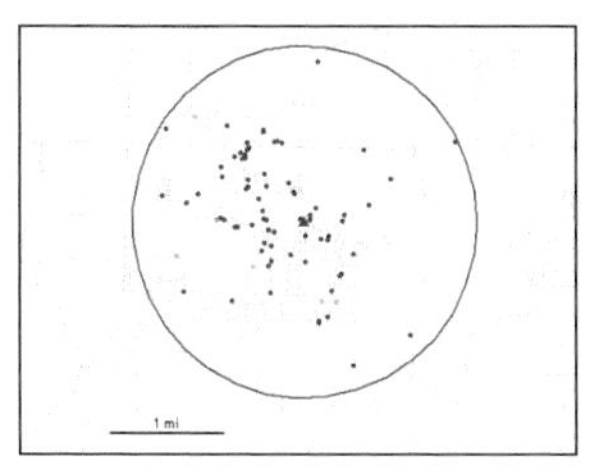

Calls to 911 within a 1.5-mile buffer around a fire station

What it's good for

This approach is good for creating a boundary or selecting features at a set distance around a source.

What you need

You need a layer containing the source feature and a layer containing the surrounding features.

DISTANCE OR COST OVER A NETWORK

You specify the source locations and a distance or travel cost along each linear feature. The GIS finds which segments of the network are within the distance or cost. You can then use the area covered by these segments to find the surrounding features near each source.

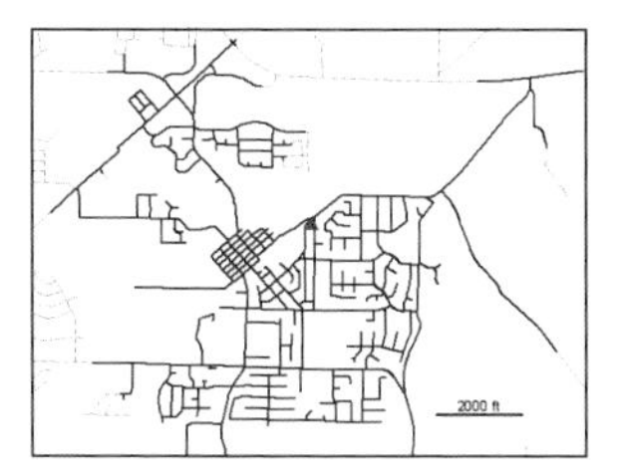

Streets within three minutes of a fire station

What it's good for

This approach is good for finding what's within a travel distance or cost of a location, over a fixed network.

What you need

You need the locations of the source features, a network layer, and–in most cases—a layer containing the surrounding features. Each segment of the network needs an attribute specifying its length or cost value.

COST OVER A SURFACE

You specify the location of the source features and a travel cost. The GIS creates a new layer showing the travel cost from each source feature.

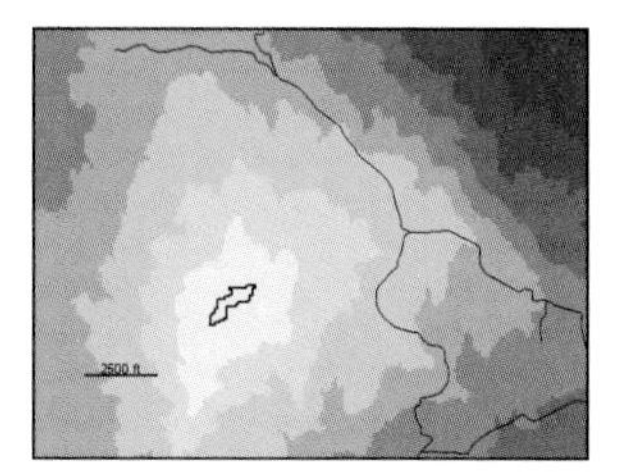

Travel cost based on slope

What it's good for

This approach is good for calculating overland travel cost.

What you need

You need a layer containing the source features and a raster layer representing the cost surface.

COMPARING METHODS

Method	Use For	Surrounding Features	Measure	Pros	Cons
Straight-line distance	defining an area of influence around a feature, and creating a boundary or selecting features within the distance	locations lines areas	distance	Relatively quick and easy	Only gives a rough approximation of travel distance
Distance or cost over a network	measuring travel over a fixed infrastructure	locations lines	distance or cost	Gives more precise travel distance/cost over a network	Requires an accurate network layer
Cost over a surface	measuring overland travel and calculating how much area is within the travel range	continuous raster surface	cost	Lets you combine several layers to measure overland travel cost	Requires some data preparation to build the cost surface

CHOOSING A METHOD

Use the following guidelines to help you choose the best method:

- Use straight-line distance if you're defining an area of influence or want a quick estimate of travel range.
- Use cost or distance over a network if you're measuring travel over a fixed infrastructure to or from a source.
- Use cost over a surface if you're measuring overland travel.

Using straight-line distance is a quick way of seeing which features are within a given distance of a source feature, and getting information about them. You can do this in several ways:

- Create a buffer to define a boundary and find what's inside it.

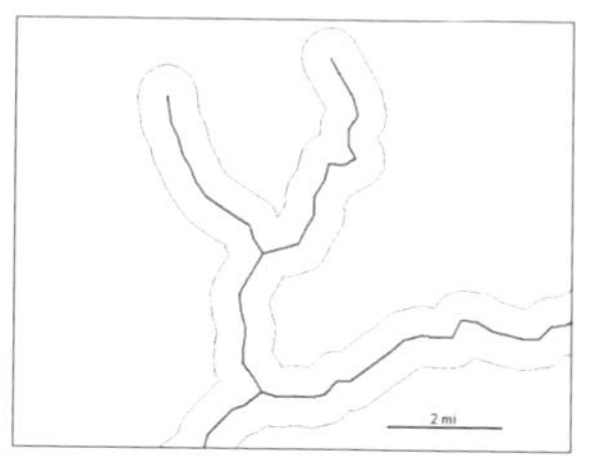

Buffer around streams

- Select features to find features within a given distance.

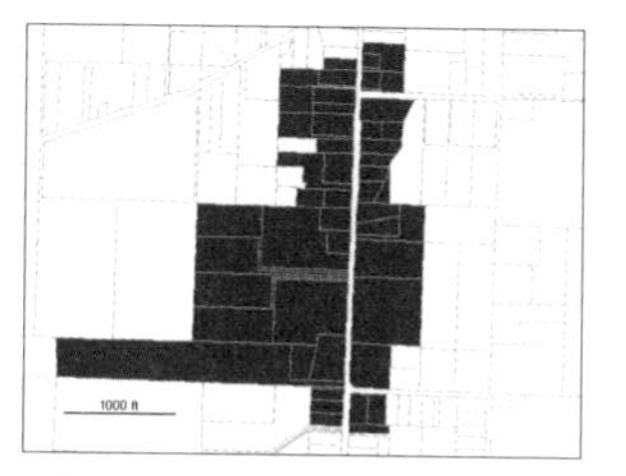

Selected parcels within 100 feet of a road

- Calculate feature-to-feature distance to find and assign distance to locations near a source.

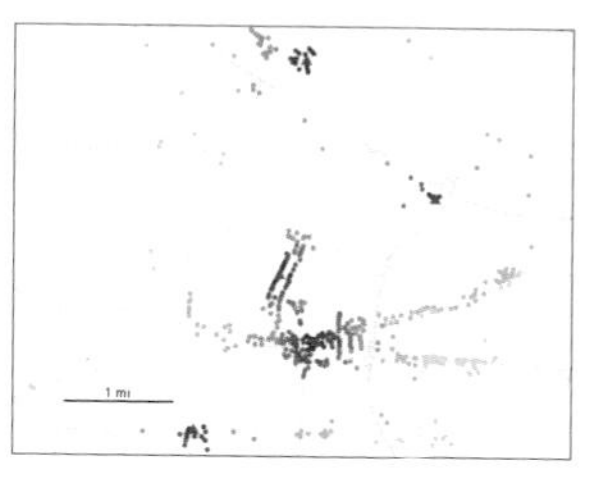

Customer locations color-coded by distance from a bank

- Create a distance surface to calculate continuous distance from a source.

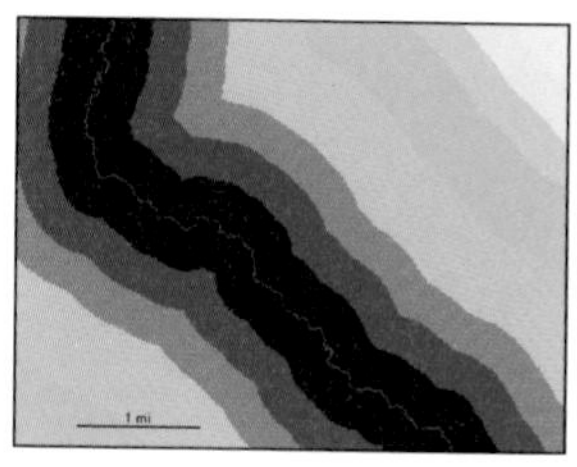

Continuous distance from a stream

With all these approaches, the GIS calculates the distance based on a straight line, or Euclidean, distance, using simple geometry. Since the GIS stores the coordinates for each point, it can calculate the distance in the x and y directions between two points, and thus the straight-line distance between them.

$$\text{distance} = \sqrt{(x_1-x_2)^2 + (y_1-y_2)^2}$$

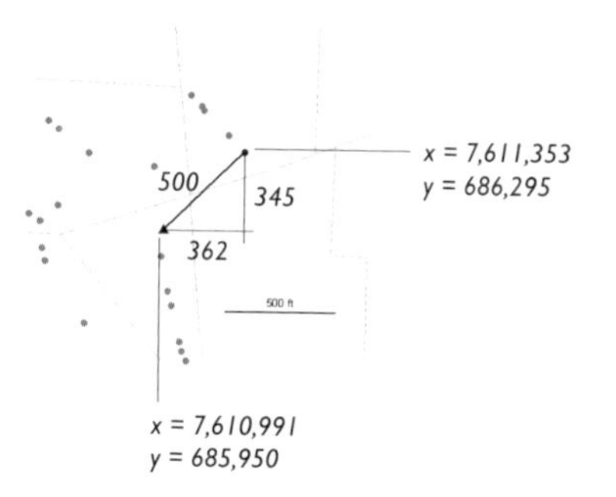

These methods are discussed in more detail in the following sections.

CREATING A BUFFER

To create a buffer, you specify the source feature and the buffer distance. The GIS draws a line around the feature at the specified distance. You can save the line as a permanent boundary, or use it temporarily to find out what or how much of something is inside the area.

For locations, the GIS draws a circle of a radius equal to the distance you specified. For linear features, the GIS draws a line around the feature at the specified distance. For areas, the GIS draws a line at the specified distance from the boundary—rather than the center—of the area.

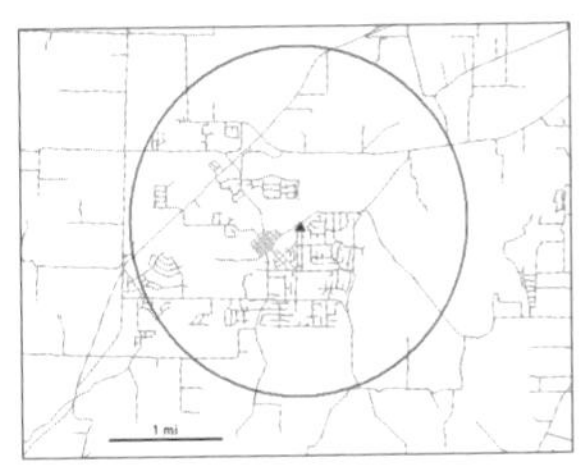

¾-mile buffer around a store

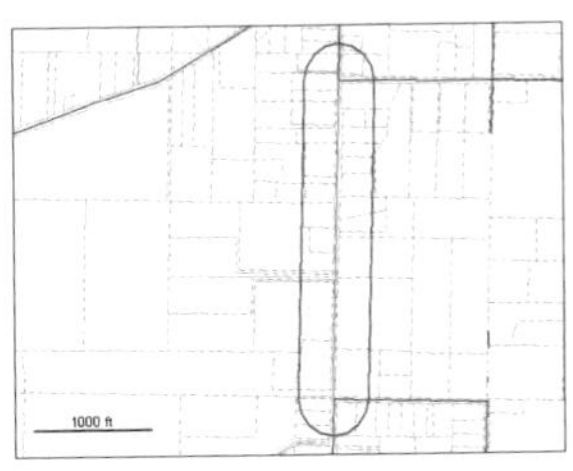

200-foot buffer around a street

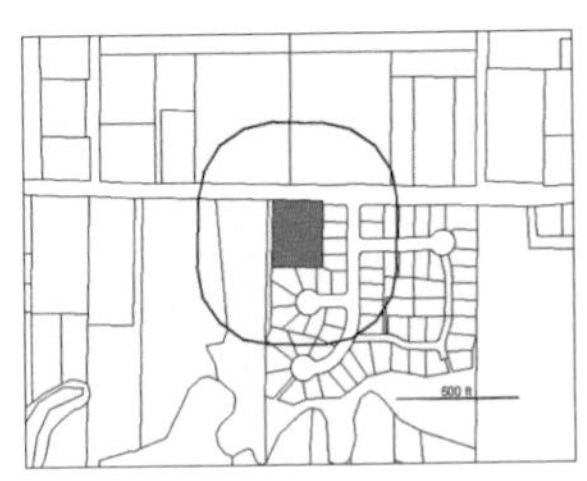

300-foot buffer around a parcel

If you have several source features, you can have the GIS buffer each source at the same distance or have it draw a variable distance buffer based on an attribute of each. For instance, if calculating a noise buffer around roads, you might specify a distance of 100 feet for highways, 50 feet for secondary streets, and 25 feet for local roads. The GIS will draw the buffer width based on the type of street.

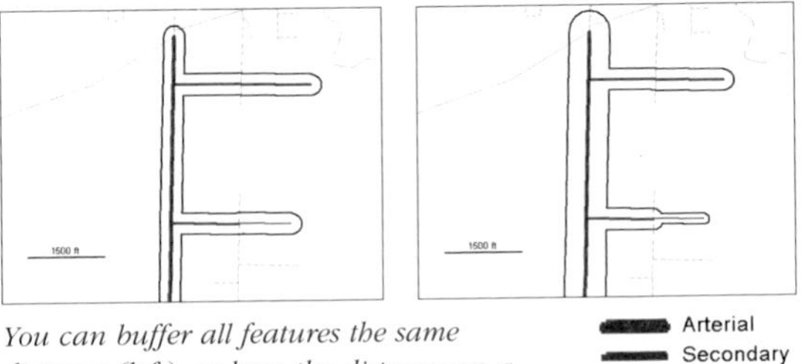

You can buffer all features the same distance (left), or base the distance on a feature attribute (right).

Arterial
Secondary
Local

You can also specify several source features and the GIS will create buffers around all of them at once. If there is overlap, you can have the GIS erase the lines where they intersect to create a single buffer area, or just let them overlap. Creating a single buffer shows you which features are near at least one source. Overlapping areas show you which features are near more than one source.

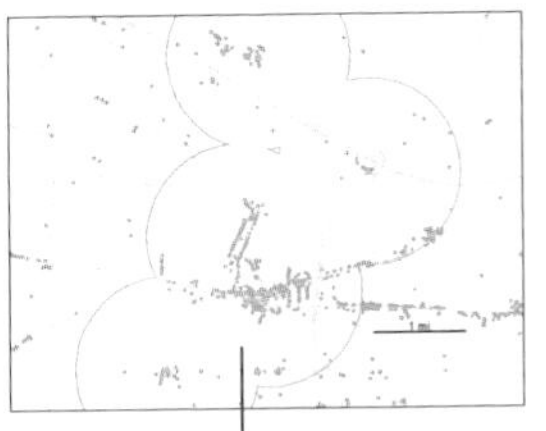

Customers inside the buffer are within ¾ of a mile of at least one bank.

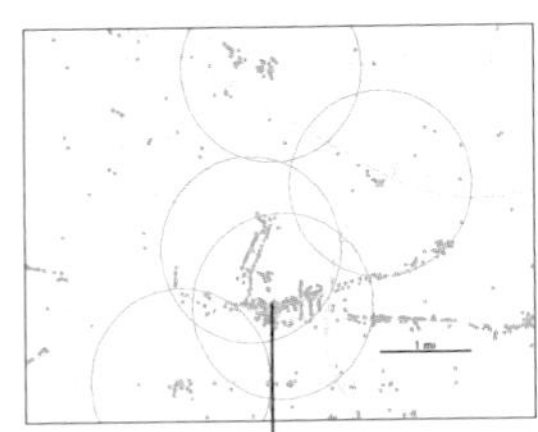

Customers in this area are within ¾ of a mile of two banks.

Getting the information

Once you've created the buffer, you can display it to see what's within the distance of the source, or you can use the buffer to select the features that fall within it. That lets you get summary information about them, such as a list or count.

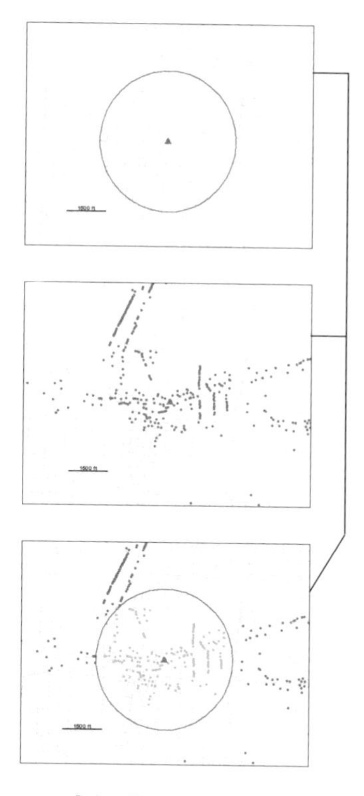

Business Name	SIC	Type
Gresham Ford	5511	Retail
Coffee People	5499	Retail
Burger King	5812	Retail
Cowtown Boots	5661	Retail
Red Robin International	5812	Retail
Redmart Inc	5712	Retail

After creating the buffer, use it to select the features inside.

If the features inside the buffer are lines or areas, you'll need to decide whether you want to include only features that fall completely inside the buffer, features partially inside, or just the portion of each feature that falls inside.

Chapter 5, 'Finding what's inside,' discusses how to get information about what's inside an area.

Finding features near several sources

If you want to find features within the distance of more than one source feature, you'll need to create separate buffers and select the features surrounding each. Otherwise, you'll only know that a surrounding feature is within the distance of at least one source—you won't know which one, or if it's within the distance of more than one source. For example, creating ¾-mile buffers around several fire stations and selecting calls to 911 within the buffers doesn't show to which station each call is nearest. You need to buffer each station separately to find the calls within ¾ of a mile of each.

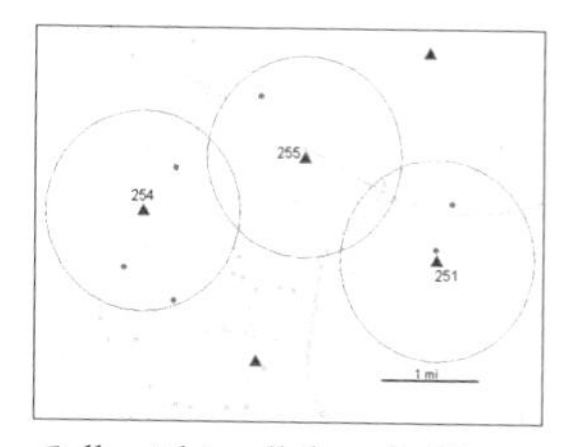

Calls within all three buffers around fire stations

Call ID	Date/Time	Type
98007009	3/21/98 20:04:40	CFIRE
98007581	3/28/98 9:50:26	CFIRE
98001957	1/20/98 1:05:50	CFIRE
98011257	5/09/98 15:00:53	CFIRE
98010414	4/29/98 23:00:59	CFIRE
98012985	5/30/98 6:16:40	CFIRE

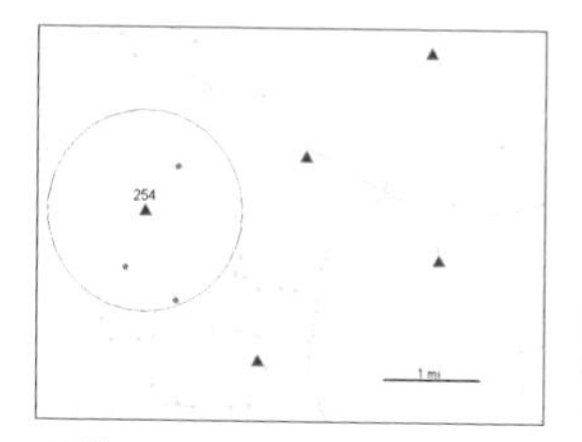

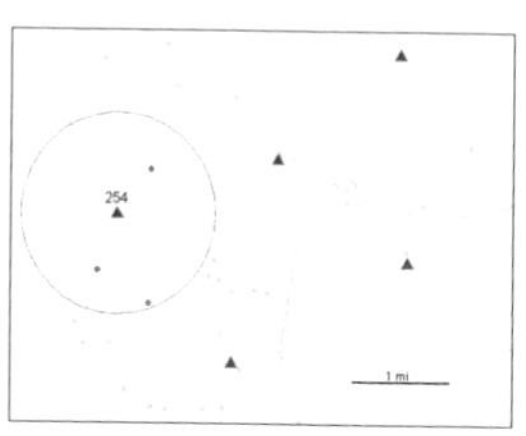

Calls near each station

Call ID	Date/Time	Type
98007009	3/21/98 20:04:40	CFIRE
98007581	3/28/98 9:50:26	CFIRE
98011257	5/09/98 15:00:53	CFIRE

Call ID	Date/Time	Type
98001957	1/20/98 1:05:50	CFIRE
98010414	4/29/98 23:00:59	CFIRE

Finding features within several distance ranges

If you want to know which features are within several distance ranges of the source as inclusive rings—such as customers within 1,000, 2,000, and 3,000 feet of a store—you have to create several separate buffers and select the surrounding features for each.

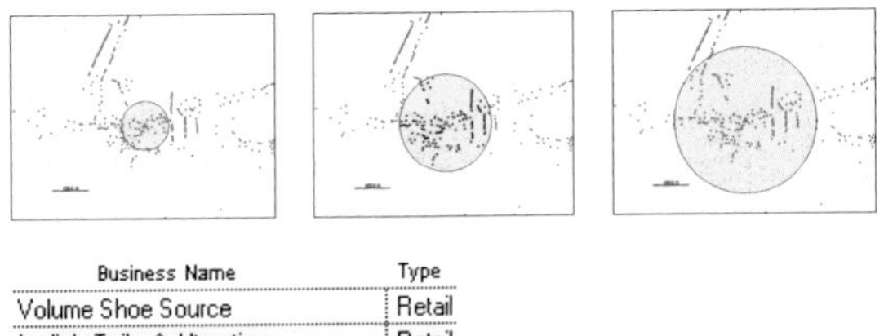

Business Name	Type
Volume Shoe Source	Retail
Lydia's Tailor & Alterations	Retail
Washco Lawn & Garden	Retail
R K's Prints & Frame Shop	Retail
Jay Jacobs	Retail
Apsara Restaurant	Retail
[illegible]	[illegible]

To find features within a distinct band of distance—for instance, customers within 1,000 feet, customers 1,000 to 2,000 feet away, and customers 2,000 to 3,000 feet away—you specify these three bands and the GIS creates them at one time. You then select the features inside the band you're interested in.

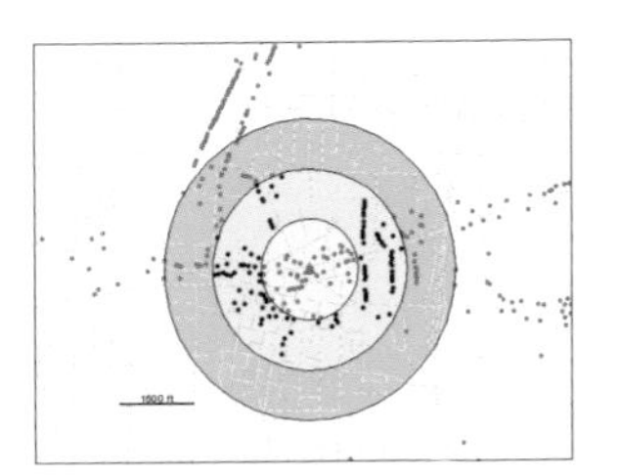

Business Name	Type	Distance Range (Ft)
Lydia's Tailor & Alterations	Retail	1000
Washco Lawn & Garden	Retail	2000
R K's Prints & Frame Shop	Retail	2000
Jay Jacobs	Retail	
Apsara Restaurant	Retail	2000
Woodstove Emporium Inc	Retail	
River Forum Sundries	Retail	
Meyers Cafe	Retail	3000
[illegible]	[illegible]	

Making a map

If you want readers to focus only on what's inside the buffer, you can show only those features. However, showing all the features can help readers see which features are outside, as well as inside, the buffer.

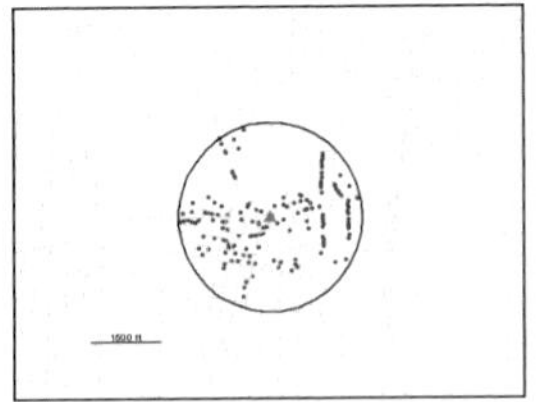

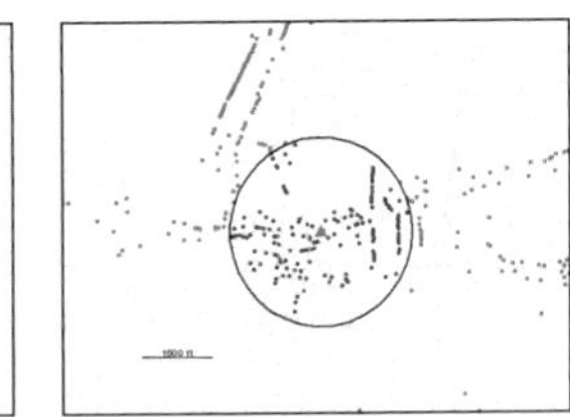

If you're displaying all surrounding features, highlighting the ones inside the buffer can help readers quickly see what's inside.

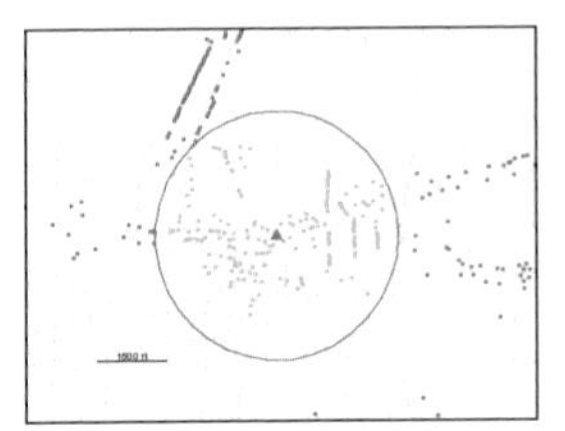

You'll also want to show features such as streets, administrative boundaries, water bodies, or other landmarks. The map should also clearly indicate what the buffer distance value is, and what it represents (for example, the area off limits to logging, or lots within 500 feet of a proposed liquor license). You can do this as text in the legend, as a label on the map, or even in the title.

SELECTING FEATURES WITHIN A DISTANCE

Using selection to find what's nearby is similar to creating a buffer. You specify the distance from the source and the GIS selects the surrounding features within the distance. The difference is the GIS doesn't create a boundary around the source features. It calculates the distance and selects the features in one step, so you don't have to use the buffer to select the features surrounding the source.

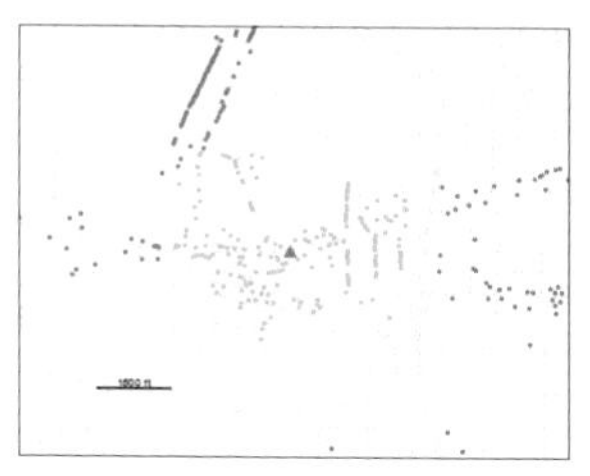

Selected customers within ¾ of a mile of a bank

Selecting features is useful if you need a summary of features near a source, and don't need to create and display an actual buffer boundary. When selecting line or area features near a source, the GIS includes features if any portion of the feature is within the specified distance.

Getting the information

Once the GIS has selected the features, you can get a list, count, or summary statistic based on an attribute.

Selecting features near several sources

If you want to find out which features are within the distance of more than one source feature, you have to select each one and tag it with a code. Otherwise, you'll only know that a surrounding feature is within the distance of at least one source—you won't know which one, or whether it's within the distance of more than one source. For example, you'd select features within 2,000 feet of bank branch 4218, and set an attribute in the data table for those features to 1. Then, you'd select features within 2,000 feet of bank branch 4220 and set another attribute for those features to 1. By then selecting customers with both codes equal to 1, you'd get a list of customers within 2,000 feet of both branches.

1 Select and tag features within the distance of the first source.

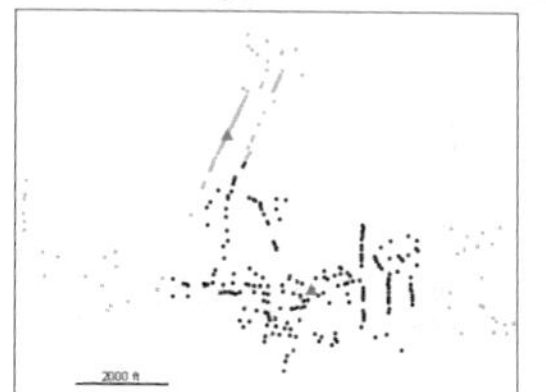

Business Name	Type	Bank #4218
Copelands Sports	Retail	1
The Little Mermaid	Retail	1
MacKenzie Roadhouse Grill	Retail	0
Spagetti Warehouse	Retail	1
Minutes Service Centers	Retail	1
Silicon Forest Computers	Retail	1
Canyon Service Center Inc	Retail	1
Peggy's Classic Cars Inc	Retail	0

2 Select and tag the features within the distance of the second source.

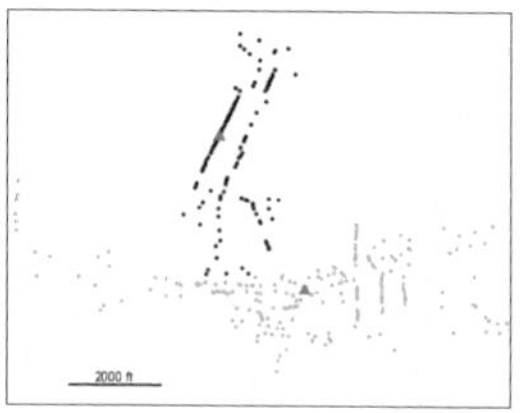

Business Name	Type	Bank #4220
Copelands Sports	Retail	0
The Little Mermaid	Retail	0
MacKenzie Roadhouse Grill	Retail	0
Spagetti Warehouse	Retail	1
Minutes Service Centers	Retail	0
Silicon Forest Computers	Retail	1
Canyon Service Center Inc	Retail	0
Peggy's Classic Cars Inc	Retail	0

3 Select features within the distance of both sources.

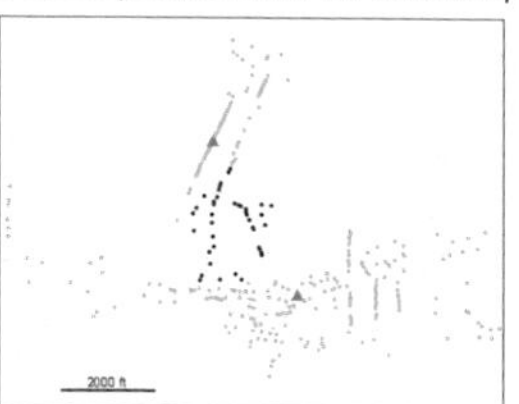

Business Name	Type	Bank #4218	Bank #4220
Copelands Sports	Retail	1	0
The Little Mermaid	Retail	1	0
MacKenzie Roadhouse Grill	Retail	0	0
Spagetti Warehouse	Retail	1	1
Minutes Service Centers	Retail	1	0
Silicon Forest Computers	Retail	1	1
Canyon Service Center Inc	Retail	1	0
Peggy's Classic Cars Inc	Retail	0	0

Selecting features within several distance ranges

If you want to know what's within several distances of the source—for example, customers within 1,000, 2,000, and 3,000 feet of a store—you perform the selection once for each distance.

If you want to find out which features are within a band of distance —for instance, customers within 1,000 feet, customers between 1,000 and 2,000 feet, and customers between 2,000 and 3,000 feet from a store—you'd select the features within each distance and tag them with a code. For example, select customers within 1,000 feet and set an attribute for those features to 1 in the data table. Then select customers within 2,000 feet and set another attribute for those features to 1. You can then select customers more than 1,000 feet but less than 2,000 feet from the store (band1000 = 0 and band2000 = 1).

1 Select and tag customers within 1,000 feet of the bank.

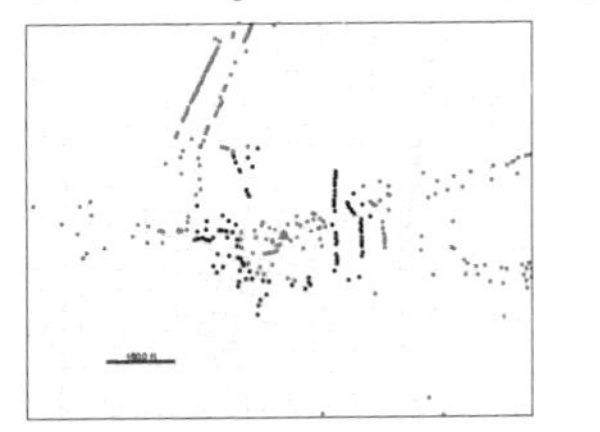

Business Name	Type	Band 1000	Band 2000
L & Z Specialties Inc	Retail	0	0
Copelands Sports	Retail	1	1
The Little Mermaid	Retail	0	1
MacKenzie Roadhouse Grill	Retail	0	0
Spagetti Warehouse	Retail	0	1
Minutes Service Centers	Retail	1	1
Silicon Forest Computers	Retail	0	0

2 Select and tag customers within 2,000 feet of the bank.

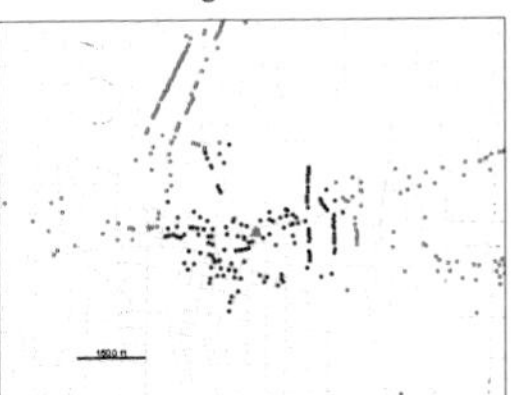

Business Name	Type	Band 1000	Band 2000
L & Z Specialties Inc	Retail	0	0
Copelands Sports	Retail	1	1
The Little Mermaid	Retail	0	1
MacKenzie Roadhouse Grill	Retail	0	0
Spagetti Warehouse	Retail	0	1
Minutes Service Centers	Retail	1	1
Silicon Forest Computers	Retail	0	0

3 Select customers more than 1,000 feet but less than 2,000 feet from the bank.

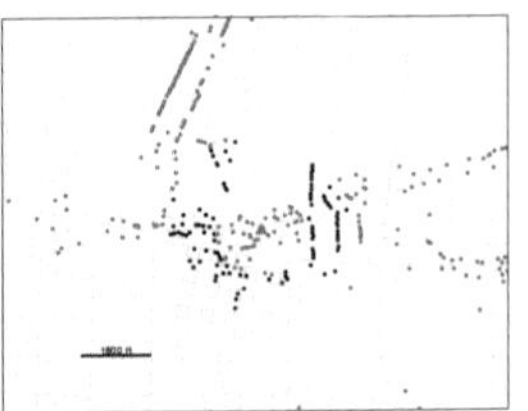

Business Name	Type	Band 1000	Band 2000
L & Z Specialties Inc	Retail	0	0
Copelands Sports	Retail	1	1
The Little Mermaid	Retail	0	1
MacKenzie Roadhouse Grill	Retail	0	0
Spagetti Warehouse	Retail	0	1
Minutes Service Centers	Retail	1	1
Silicon Forest Computers	Retail	0	0

Making a map

To display what's near the source feature, you simply draw the source and the selected surrounding locations. However, you'll probably want to draw all the locations in the study area, and then draw the ones within the specified distance using a different color or symbol to distinguish them. That lets you know where the locations outside the distance are, as well as the ones inside.

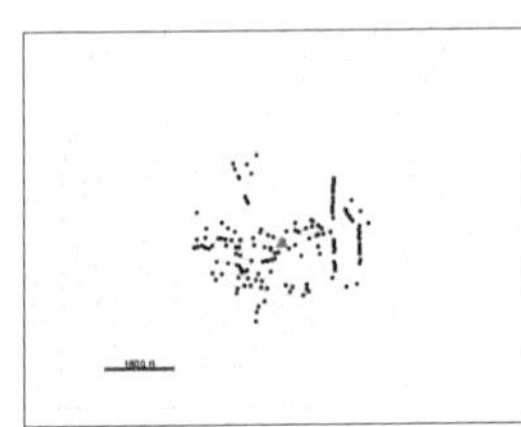

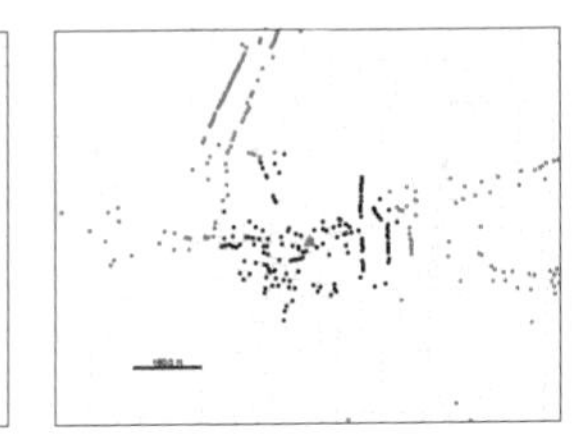

FEATURE TO FEATURE

If you're finding individual locations near a source feature, you can have the GIS calculate the actual distance between each location and the closest source. This is useful if you need to know exactly how far each location actually is from the source, rather than just whether it's within a given distance. If you're finding the distance to a linear feature, such as how far eagle nests are from a river, the GIS calculates the distance to the closest point on the line.

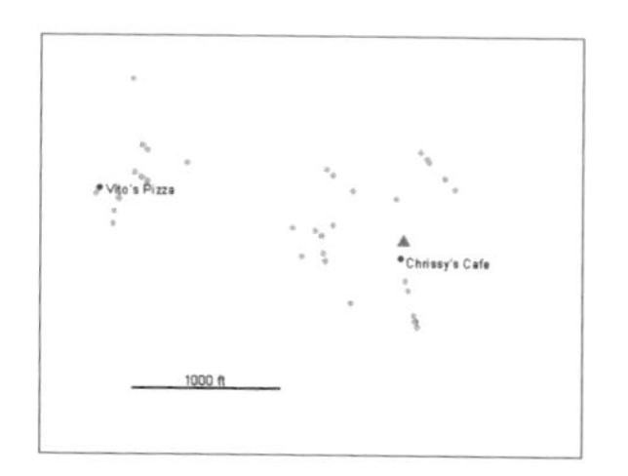

Business Name	SIC	Type	Distance (Ft)
Vitos Pizza	5812	Retail	2091.394
Chrissys Cafe	5812	Retail	119.111
Premier Technologies Inc	5734	Retail	2126.888

You can also have the GIS find the distance between each location and several source features. This is useful if:

- You want to see which areas are near more than one source and which areas are near only one. For example, owners of a fast food chain might want to see which areas have many customers near several restaurants.
- You need to know the second or third closest source for each location. For example, you might want to know the nearest and next nearest hospitals to a given location.
- You want to compare distance to other factors. For example, you could compare each customer's distance from the store to the number of store visits, and generalize this for all customers to see the relationship of distance to store visits.

What you get

If you're calculating the distance between each surrounding location and the nearest source, the distance for each location is added to the data table for the surrounding locations, along with an identifier for the nearest source.

Business Name	SIC	Bank #	Distance (Ft)
Spanky's Burger Express	581	3	1252.373
Copeland Lumber Yards Inc	521	3	1379.105
Spunky's Hamburgers	581	3	1297.968
El Torito Restaurant	581	3	1685.213
Marion's Carpets Inc	571	3	1442.558
Sherwin Williams Paint	523	3	1838.423
Godfathers Pizza	581	3	1475.071
GDR's Golf USA	594	3	1420.045
The Printer Place	573	3	1579.779

If you're calculating the distance between each location and several sources, you get a new table listing, for each location, an identifier for the source and the distance to each.

Business Name	Bank Branch	Distance (Ft)
Spanky's Burger Express	3	1252.373
Copeland Lumber Yards Inc	3	1379.105
Copeland Lumber Yards Inc	4	2315.988
Spunky's Hamburgers	3	1297.968
El Torito Restaurant	3	1685.213
El Torito Restaurant	4	2021.815
Marion's Carpets Inc	3	1442.558
Marion's Carpets Inc	4	[illegible]

Specifying a maximum distance

If you're calculating the distance to more than one source, you can specify a maximum distance within which locations will be included. For example, if you're interested in getting information only about customers living within 5 miles of your store, you'd specify a maximum distance of 5 miles, and the GIS would not include any customers beyond this limit.

Specifying a maximum distance is usually a good idea—otherwise, the GIS will create a list of the distance between each source and every location in the study area. For example, you'd get a list showing the distance for each customer to each store. You set the maximum distance based on your own knowledge of how people or things behave—for instance, a distance beyond which people generally won't travel to shop at a particular type of store.

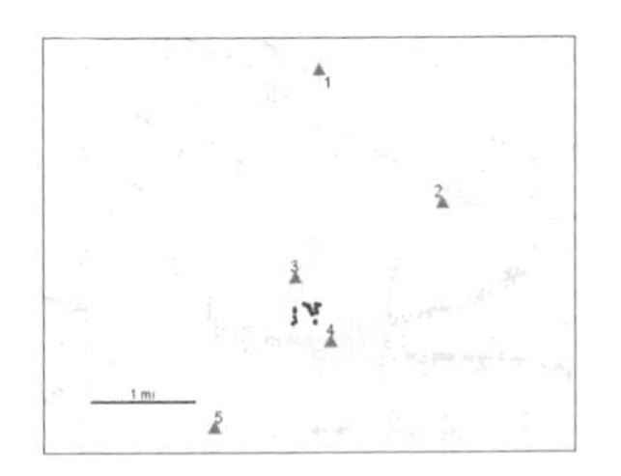

Business Name	Bank Branch	Distance (Ft)
Spanky's Burger Express	3	1252.373
Copeland Lumber Yards Inc	3	1379.105
Copeland Lumber Yards Inc	4	2315.988
Spunky's Hamburgers	3	1297.968
El Torito Restaurant	3	1685.213
El Torito Restaurant	4	2021.815
Marion's Carpets Inc	3	1442.558
[illegible]	4	[illegible]

Getting the information

Since each location is tagged with its distance from the source, it's easy to map what's within several distances. For example, you may want to know which customers are within 5 and 10 miles of a store. To find this out, you simply classify the locations based on their distance from the source. You could also select those features and get a list of them.

This method also lets you calculate statistical summaries based on the assigned distance, such as the average distance businesses are from a bank branch, or the median distance eagle nests are from a river. You specify the field containing the code of each source, and the attribute you want to summarize. The GIS does the calculation and creates a table or chart showing the value for each source location.

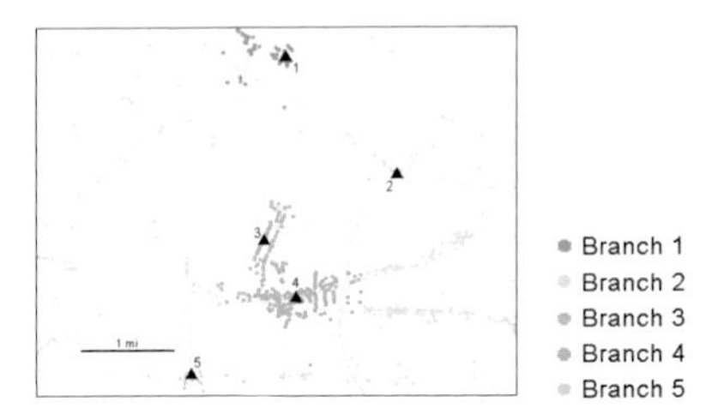

Business Name	Bank Branch	Distance (Ft)
Cent Wise Drug & Hallmark Shop	2	98.956
Century Pharmacy-Sunset	1	2506.843
Charlotte's Weddings & More	0	0.000
Chelsea Audio-Video	4	1818.969
Chicken On The Run	0	0.000
[illegible]	4	[illegible]

Bank Branch	# of Businesses	Avg. Distance (Ft)
1	56	1650.2130
2	16	767.2087
3	124	995.0087
4	206	1558.0573
5	26	1110.3992

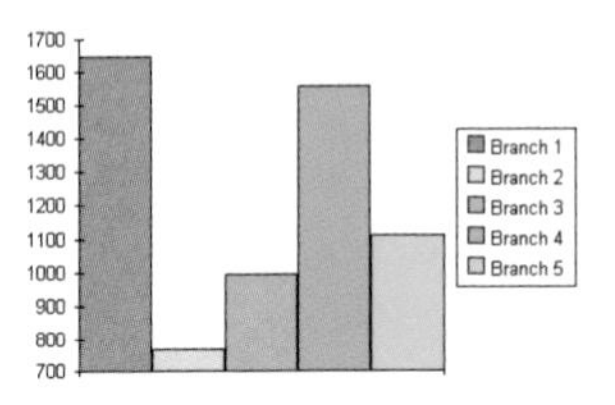

Average distance of retail businesses from the nearest bank branch

Making a map

You have several options for creating a map based on point-to-point distance:

- Map surrounding locations color-coded by distance from the source
- Map surrounding locations color-coded by the closest source
- Create a spider diagram
- Map source features using graduated point symbols

Map surrounding locations color-coded by distance

This option shows you how near locations are to a source feature. You simply classify the distance values into ranges, as described in chapter 3, 'Mapping the most and least.' Use several hues to make it easier to distinguish the ranges.

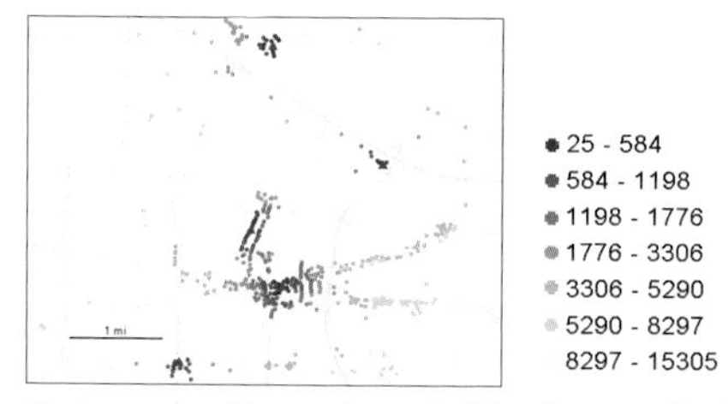

Customer locations color-coded by distance (in feet) from the nearest bank branch

Map surrounding locations color-coded by source

This option shows you to which source feature each location is nearest. It also helps you see the area served by each source. Since a code for the source is stored for each location, you simply assign a color to each code, and the GIS draws each location with the appropriate color.

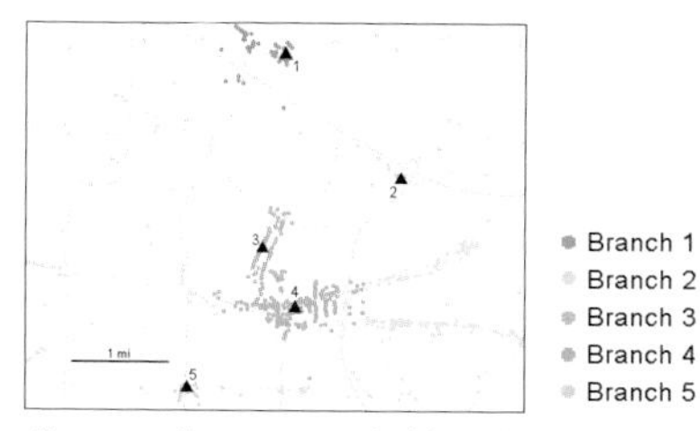

Customer locations coded by which bank branch they're nearest.

Create a spider diagram

You can have the GIS draw a line between each location and its nearest source. This is often called a spider diagram. If a location is near two or more sources, the GIS draws a line to each. You can draw the lines in different colors to make it easy to see which locations are associated with each source. Spider diagrams are especially useful for comparing the varying patterns between several source features, such as how far—and in which direction—locations are from a source, which source features have more locations near them, and which locations are near two (or more) sources.

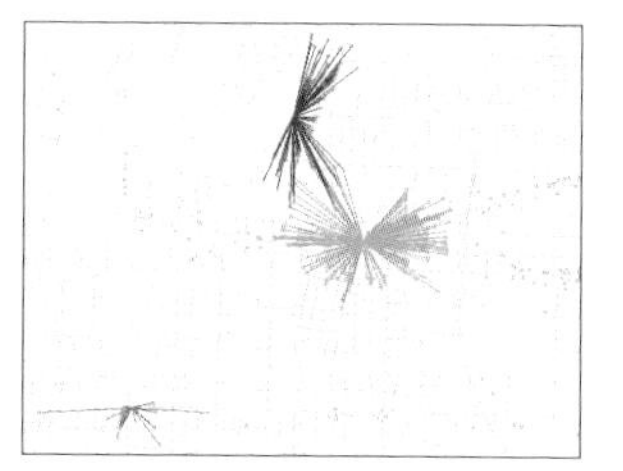

Customers within ¾ of a mile of a bank

Map source features using graduated point symbols

Graduated symbols are useful for comparing source features based on a quantity. The symbols indicate the number of locations near each source feature (for example, the number of businesses near each bank), or some value based on the surrounding features (for example, the number of employees at the businesses near each bank). You can also combine this with other methods of drawing the surrounding locations, such as color coding by distance or by source.

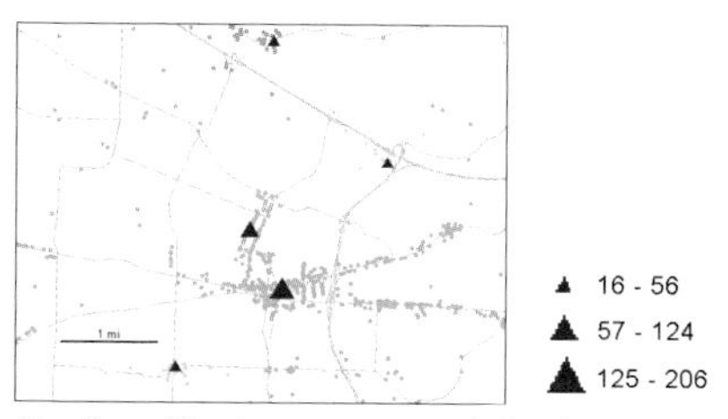

Number of businesses near each bank

CREATING A DISTANCE SURFACE

Using this method, you create a raster layer of continuous distance from the source. You can use the distance layer to create buffers at specific distances, then assign distance to individual features surrounding the source or find how much of a continuous feature, such as soil or vegetation, is near a source feature.

You specify the layer containing the source features, and the GIS creates a new raster layer by calculating the distance from each cell to the nearest source.

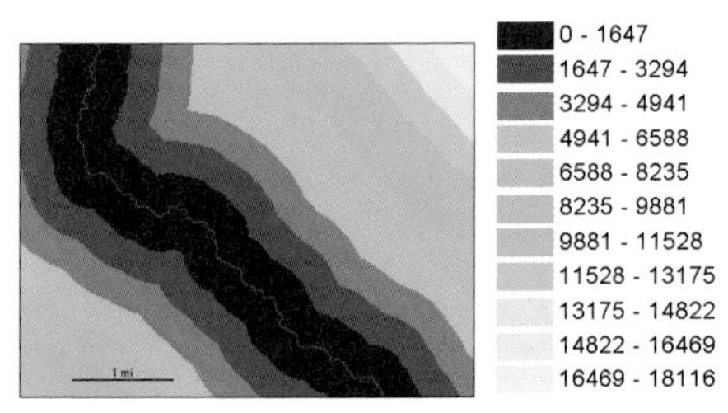

Continuous distance from a stream, in feet

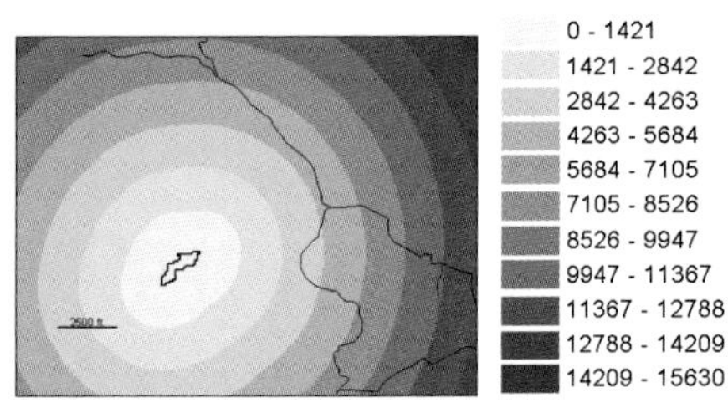

Continuous distance from a timber stand, in feet

Creating distance ranges

Each cell potentially has a unique value. You display the values using graduated colors—either as a continuous range or grouped into classes—so you can see the patterns. ArcInfo and ArcView GIS create default displays. If you just want to display the distance layer with other features to see what's within the distance, you can use the default or specify your own color ranges or classes.

Summarizing what's within the distance

You can summarize either discrete features or continuous data within the distance.

To summarize locations, such as nests near streams, you assign the distance to each feature, based on the cell it falls within. You can then summarize the features by distance. For more information, see the section 'Feature to feature.'

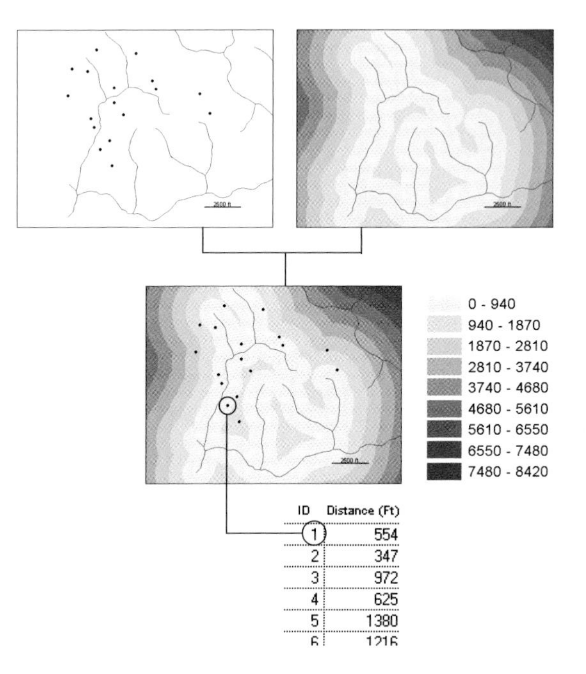

To summarize continuous features—for example, the amount of each land cover type within 500 feet of a stream—you select cells with a value less than or equal to 500 to create a new layer. The layer shows which cells are within the 500-foot buffer. Then overlay the distance layer and the layer containing the continuous features.

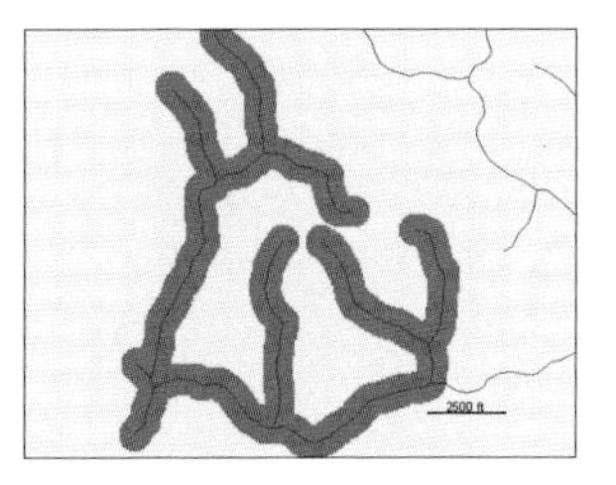

To create multiple distance buffers, you reclassify the layer into distance ranges, for example, 0–500, 500–1000, and 1000–1500 feet.

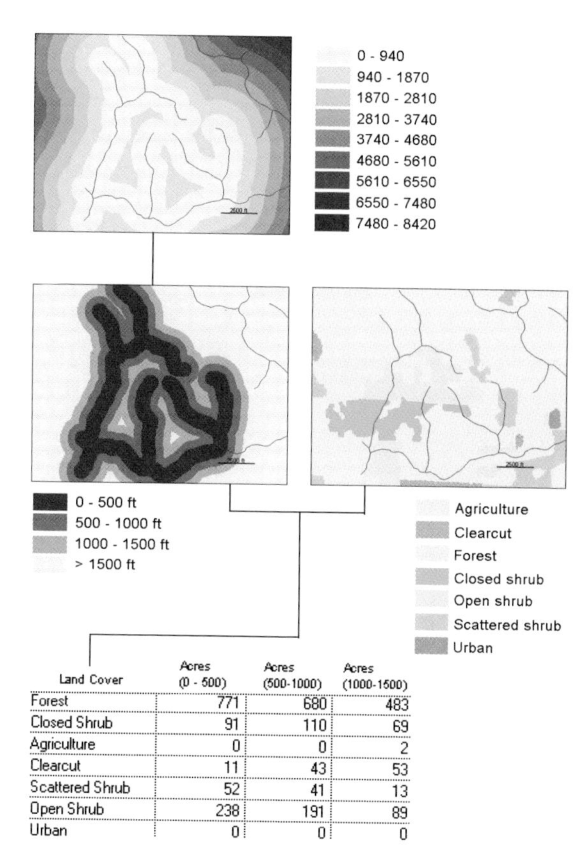

Land Cover	Acres (0 - 500)	Acres (500-1000)	Acres (1000-1500)
Forest	771	680	483
Closed Shrub	91	110	69
Agriculture	0	0	2
Clearcut	11	43	53
Scattered Shrub	52	41	13
Open Shrub	238	191	89
Urban	0	0	0

This method is also often used to create input to site selection or suitability models, where distance from a source is a factor. The distance surface is reclassified into ranges by assigning relative values and then combined with other layers to assign an overall rank to each cell. For instance, distance from streams may be one criterion when you're identifying and ranking areas based on likelihood of being good deer habitat.

Chapter 5, 'Finding what's inside,' discusses how to find what's inside an area, including areas created based on distance from a source feature.

Specifying a maximum distance

You can limit the area for which the GIS calculates distance by specifying a maximum distance. Any cells beyond the distance you specify will not be assigned a value. If you don't specify a maximum distance, the GIS will calculate a value for all cells in the study area, no matter how far from the source features.

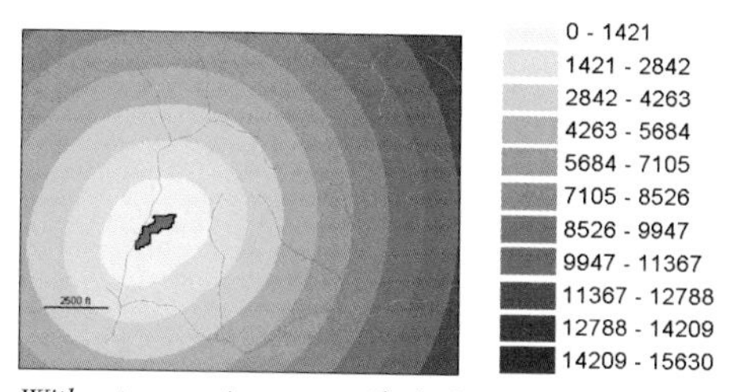

Without a maximum specified, distance is calculated for all cells.

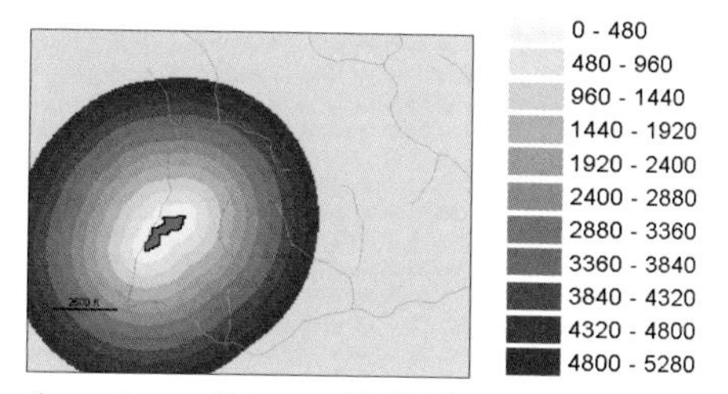

A maximum distance of 5,280 feet was specified.

Specifying multiple source features

When you create a distance surface, the values are calculated from each cell to the nearest source. If you want to find out which areas are within a given distance of more than one source, you'll need to create a separate input layer for each source and create the distance layer for each. You'll then compare the outputs by selecting the cells that are within the distance on both surfaces. For example, to find the area within 1,200 feet of two streams, you'd create a distance layer for each stream and select the cells on each that have a value of less than 1,200. The resulting layer will have only those cells that are within 1,200 feet of both streams.

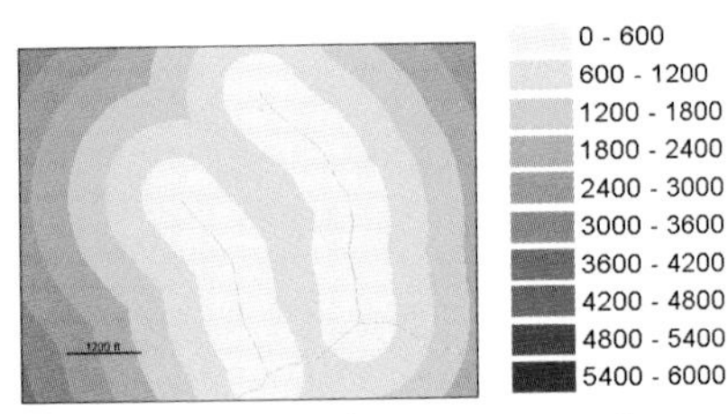

Calculating distance from two streams shows only the distance from at least one of them.

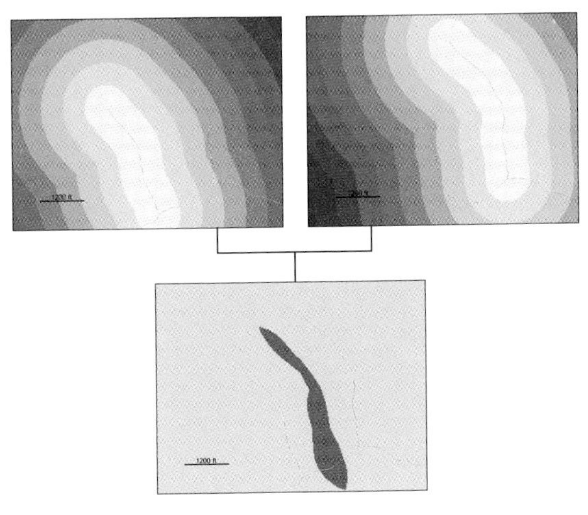

Area within 1,200 feet of both streams

Making a map

If you're mapping discrete features, you can simply draw them on top of the distance surface. The distance surface is displayed using graduated colors. If you specify more than six or seven classes, use two or three hues to help distinguish the classes. You can also use a continuous blend of colors covering several hues. Using classes, you can see the value for any location (within a range). Using a continuous blend is good for showing how values change across the surface, but makes it difficult to see the actual value at any particular spot. You'll also want to show the source features in a contrasting color.

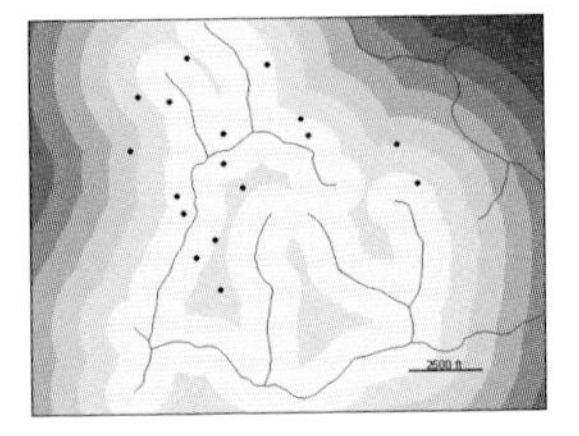

Continuous data stored as a raster can't be shown on top of the distance surface. To show what's within the distance, combine the reclassified distance surface with the surrounding features as described in the earlier section, 'Summarizing what's within the distance.' You can then display the resulting layer with the cells inside color-coded by value, and cells outside in a neutral color. You can also show all cells color-coded by their value, and highlight the ones inside the distance.

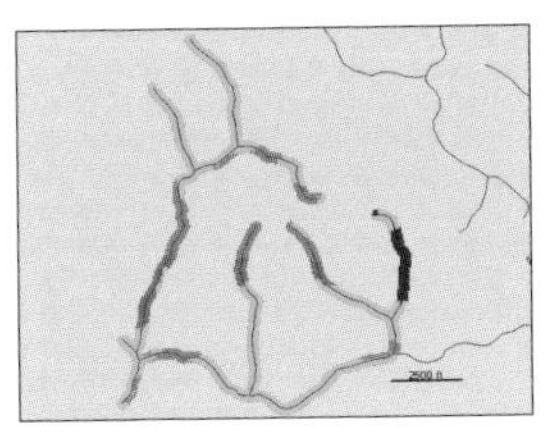

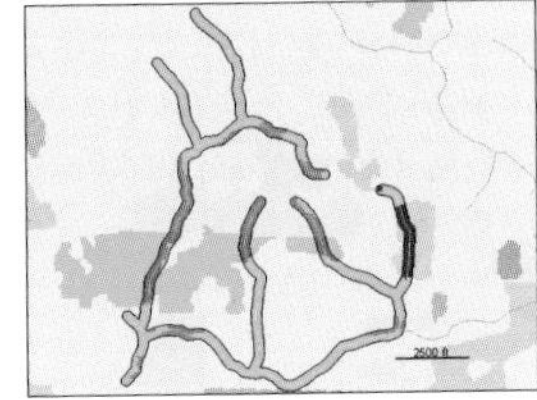

Forest
Closed shrub
Scattered Shrub
Open shrub

In this method, the GIS identifies all the lines in a network, such as streets or pipelines, within a given distance, time, or cost of a source location. Source locations in networks are often termed "centers," as they usually represent centers to or from which people, goods, or services travel. You can then find the surrounding features along—or within—the area covered by those lines. You may be interested in which lines themselves are near the center, such as the streets within three minutes of a fire station.

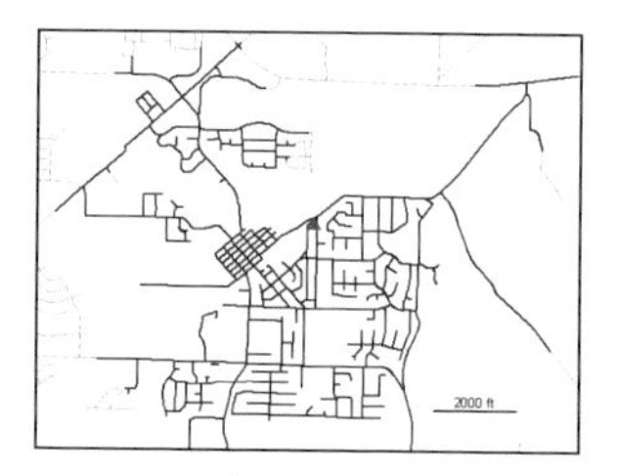

Or, you may want to know how much or how many of something is near the center, such as the number of customers within a 10-minute drive of a store.

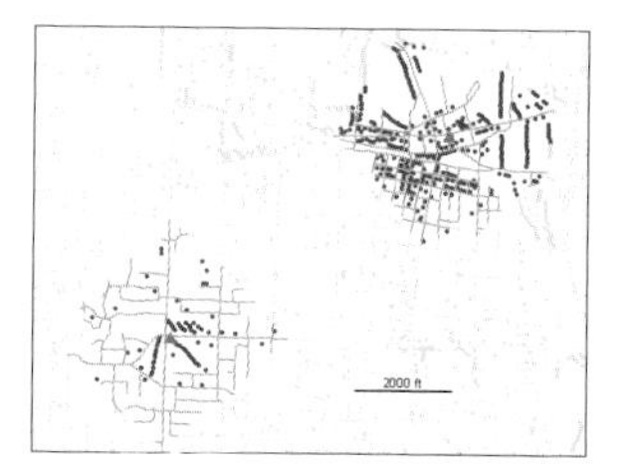

SPECIFYING THE NETWORK LAYER

A geometric network is composed of edges (lines), junctions, and turns. Junctions are the points where edges meet. Turns are used to specify the cost to travel through a junction. The GIS knows which edges are connected. To get accurate results, you should make sure your network has:

- Edges that are in the right place
- Edges that actually exist
- Edges that connect to other segments accurately
- The correct attributes for each edge

WHAT THE GIS DOES

The GIS starts at the center you specify and checks the distance to each nearest junction along the network. If the distance is less than the maximum value you specified, it tags the edge with the code for that center. It then goes to the junction at the end of one of the tagged edges and checks the distance to all the nearest junctions from there. It adds that distance to the previous distance—from the center to the current junction—to get the total distance for each. It continues the process from each junction it finds within the distance, searching outward in all directions and calculating the cumulative distance until it reaches the maximum distance you specified.

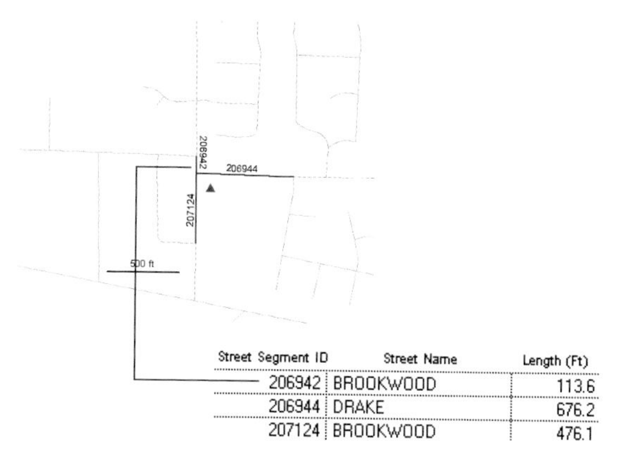

Street Segment ID	Street Name	Length (Ft)
206942	BROOKWOOD	113.6
206944	DRAKE	676.2
207124	BROOKWOOD	476.1

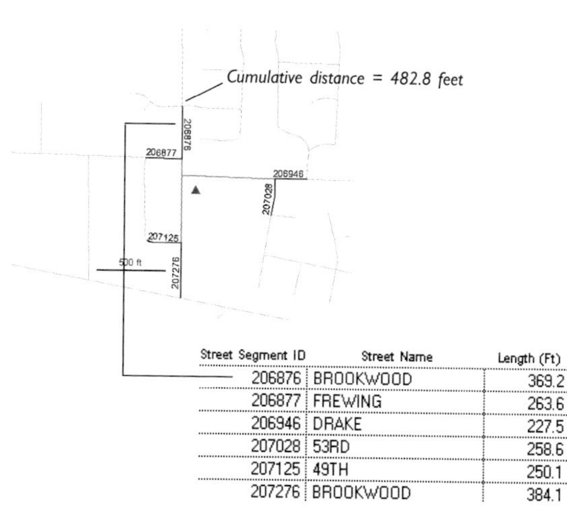

Street Segment ID	Street Name	Length (Ft)
206876	BROOKWOOD	369.2
206877	FREWING	263.6
206946	DRAKE	227.5
207028	53RD	258.6
207125	49TH	250.1
207276	BROOKWOOD	384.1

ASSIGNING STREET SEGMENTS TO CENTERS

Street networks are commonly used for finding what's nearby. These networks are composed of street segments (edges), intersections (junctions), and turns. Each street segment is tagged with a measure of the cost to travel between the center and surrounding locations. This is often termed an "impedance" value. The most common measures of impedance are distance, time, and money—for example, the per-mile cost for a delivery truck, based on labor, fuel, and maintenance.

Using distance

You specify the location of the center and the maximum distance, and the GIS assigns segments to the nearest center within the distance. The data table for the network layer includes a field containing the length of each segment, so you don't have to add this attribute.

Using cost

To find what's nearby using time or some other cost, you need to tag each street segment with its cost. One way to do this is to calculate a per-unit cost and multiply it by the length of each segment to determine the cost per segment. The per-unit cost might be a constant that you've calculated based on external information—for example, labor and fuel costs of twenty cents per mile for a delivery van. Or, the cost might be based on the street type—for example, you might assign highways a travel speed of 50 miles per hour, and residential streets a speed of 25 miles per hour.

Travel time is one of the most common costs. You can calculate a value for travel time in several ways. One method is to actually measure average travel times for each street segment or block. If you don't have travel times, but do have the speed limit stored for each street segment, you can calculate an approximate travel time by multiplying the segment length by the speed limit.

For example, if you're calculating travel time in minutes, speed limit is measured in miles per hour, and street length is measured in feet, you'd calculate:

```
minutes = length / ((mph * 5280) / 60)
```

Street ID	Name	Length (Ft)	Speed Limit (MPH)	Travel Time (Min)
219230	ARTHUR	159.675	15	0.12
219209	ROCHELLE	168.513	15	0.13
203991	14TH	431.834	25	0.20
203992	MAIN	631.283	35	0.20
218303	CORNELIUS PASS	1321.652	45	0.33
224560	QUATAMA	1597.035	35	0.52
219208	ROCHELLE	268.405	15	0.20
221358	LEONARD	125.641	15	0.10

If you don't have the speed limit for each segment, you can use the street type as a surrogate. For example, you might assign all major streets a travel speed of 40 miles per hour and all residential streets a travel speed of 25 miles per hour. You can then calculate travel times for each street segment as shown above.

You calculate monetary costs by multiplying the length of the segment by the travel cost. For example, if segment length is measured in feet, and travel cost is in cents per mile, the cost to travel each segment would be calculated as:

```
cents =
length * (cost per mile / 5280)
```

SETTING TRAVEL PARAMETERS

In addition to specifying the cost for individual segments, you can specify the cost for turns from one segment onto another, or for stops at an intersection. You can also limit which segments on the network can be traveled, and in which direction.

Specifying turns and stops

You use turns and stops most often when calculating travel time. For instance, you can specify that a right turn at a particular intersection takes an average of three seconds, while a left turn takes seven seconds. Or, that a stop for a stop sign may average three seconds, while a stop at a traffic light averages 30 seconds.

Whether or not you decide to use turns and stops depends on how exact your analysis has to be. It's more important, for example, to include turns and stops when analyzing which streets are within three minutes of a fire station than which shoppers are within a 15-minute drive of a mall.

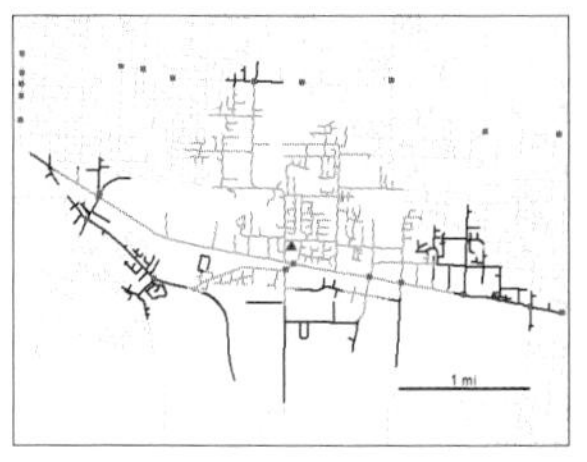

Streets in blue are within three minutes of a fire station without stops and turns, but more than three minutes when turns and stops are included.

To assign cost to a turn or stop, you have to create a turntable. A turntable is a data table that lists the junctions for which you want to specify a cost. If a junction isn't listed, the GIS assumes it has no additional cost associated with it. The turntable lists the numeric identifier of the junction, the identifier of the "from" segment in the network layer, the identifier of the "to" segment, and the cost value for the turn or stop (for example, 3 seconds, or 5 cents).

As the GIS assigns segments, it finds the ID of each junction it comes to, looks in the turntable to see whether the junction's listed, and, if so, finds the specific turn (the "to" and "from" segments) and the cost for that turn. It then adds the cost to the running total. A stop is treated in the same way as a turn.

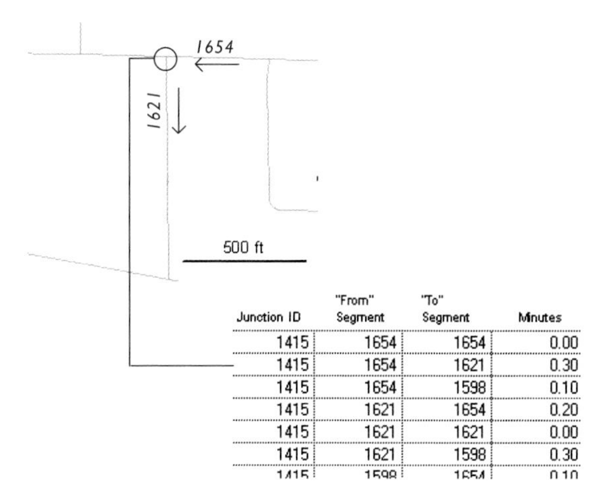

Junction ID	"From" Segment	"To" Segment	Minutes
1415	1654	1654	0.00
1415	1654	1621	0.30
1415	1654	1598	0.10
1415	1621	1654	0.20
1415	1621	1621	0.00
1415	1621	1598	0.30
1415	1598	1654	0.10

You can also specify the direction of travel, such as: one-way streets; closed segments (like streets closed for repairs); or prohibited turns, such as an intersection with no left turns allowed. The way in which you specify these limits depends on the GIS software you're using. In general, you assign a code to each segment or turn indicating whether travel is allowed.

SPECIFYING MORE THAN ONE CENTER

If you have more than one center, the GIS assigns segments to each concurrently. You can set the maximum distance or cost to be the same for each center, or specify a different distance for each. For example, distribution centers in rural areas may have a higher maximum travel cost than ones in urban areas, because the trucks have to travel farther.

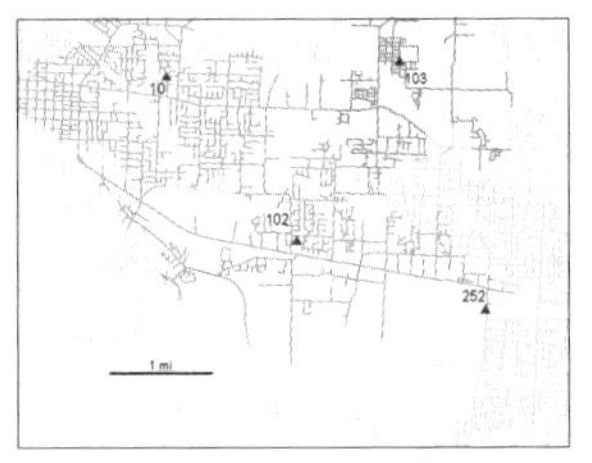

Streets are assigned to the nearest center.

By mapping several centers, you can see which areas are not near a center and which are near several. You can also see which centers have many surrounding locations—indicating high demand—and which have few.

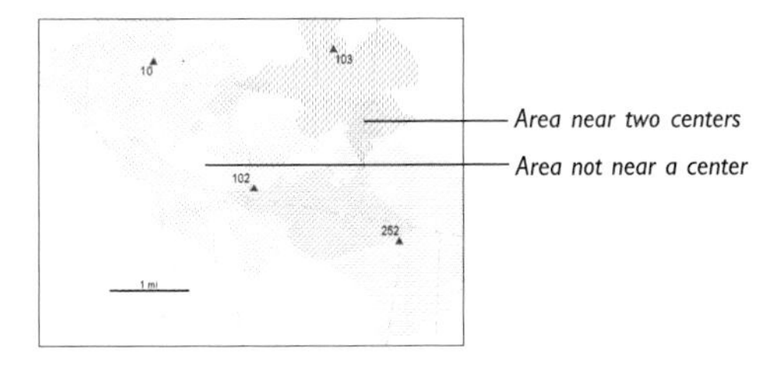

SELECTING THE SURROUNDING FEATURES

Once the GIS has identified all the segments within the distance or cost of a center, you can find out what's within the area covered by those segments. You can either create a boundary enclosing the segments and overlay that on another layer containing the surrounding features, or have the GIS sum a value associated with each segment as it searches outward from the center along the network.

Create a boundary if:

- You want a list of individual locations.
- You need a count of locations in the area covered by the selected segments.
- You have data summarized by area—for example, you want to total the number of households per census block to find out how many households are within a 15-minute drive of a recycling center.
- You need a list, count, or amount for linear features or areas—for instance, the total length of salmon streams within a half-hour drive of the town.

Sum as you go if:

- You want a precise count of locations along the network segments, or the total amount of an attribute value for those locations.
- You don't need a list of individual locations.

Using a boundary

You can create the boundary by manually drawing a line around the selected segments, or you can have the GIS create the boundary.

Drawing the boundary manually gives you more flexibility—you can use the selected segments as a guide, and include or exclude areas based on other considerations such as the locations of administrative boundaries or other features.

The GIS can draw either a compact or general boundary. A general boundary connects the farthest reaches of the selected segments, while a compact boundary outlines the selected segments. The GIS can create a general boundary much faster than a compact boundary. A general boundary might be sufficient for finding how many people live roughly within a mile of a library, along city streets, while a compact boundary would be required to find out which houses are within a three-minute drive of a fire station.

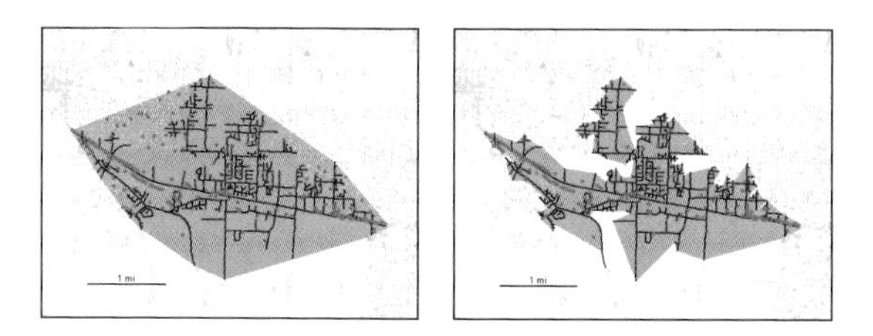

Once you've created the boundary, you find what's inside by using the boundary to select surrounding features, or overlaying it with the surrounding features. Chapter 5, 'Finding what's inside,' describes how to do this.

To find out what's within several distance or cost values, you can have the GIS create distance bands. For example, you could find customers within 0–1 mile, 1–2 miles, and 2–3 miles of a store. The GIS creates all three bands in one operation.

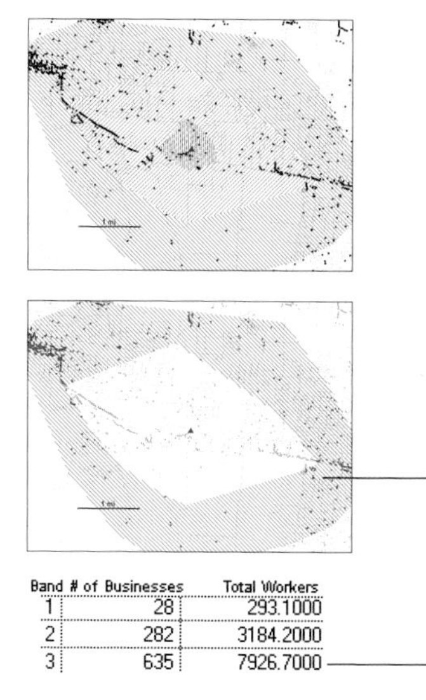

Band	# of Businesses	Total Workers
1	28	293.1000
2	282	3184.2000
3	635	7926.7000

If you want to find features within distance or cost rings—for example, customers within 0–1 mile, 0–2 miles, and 0–3 miles of a store—you have to assign segments separately for each range. You can then select what's within each area and get a list or summary.

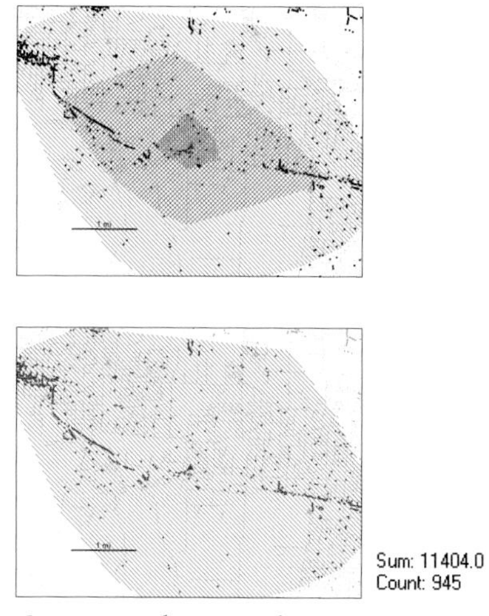

Customers within 0–3 miles.

Summing as you go

When you use this method, the GIS sums counts or amounts as it searches outward from the center along the network. To do this, you have to tag each segment with the value for that attribute. For example, you could find out how many people work on each street segment by summing the number of workers at each business, based on address. You'd then store this number for each segment in the street layer's data table. As the GIS assigns street segments to the center, it keeps a running total of the number of workers. When it reaches the maximum distance you specified, it assigns the final total to a field in the center's data table.

You can also specify that the GIS stop assigning streets once it reaches a maximum count or amount. For example, you could assign students to an elementary school by assigning streets until a total of four hundred students is reached. All the students living on those streets would be assigned to the school.

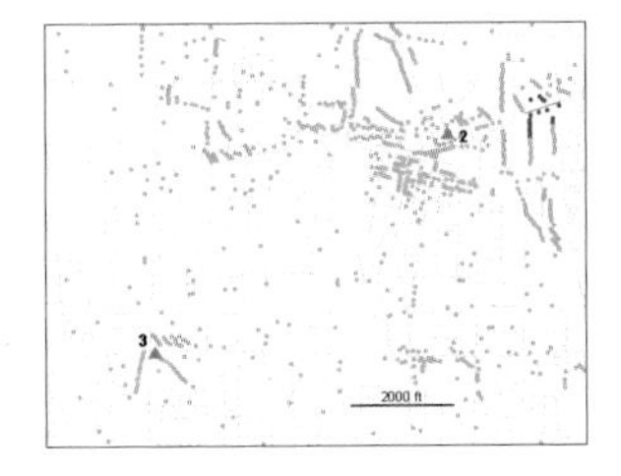

Business Name	Street	Street ID	# of Workers
Beavertons Carr Chevrolet-Geo-Subaru	Canyon	493	204
Hot Spot Fireplace & Patio Shop	Canyon	493	6
California Wholesale	Tualatin Valley	477	3
Petco	117th	497	6
The Grapery	Cedar Hills	484	6
Northwest Tub Repair	Cedar Hills	485	1

1 Sum workers by street segment.

Street ID	# of Workers
491	44
493	513
494	3
495	1
496	13

2 Assign number of workers to each street segment in the street data table.

Street ID	Street	# of Workers
493	CANYON	513
494	BEAVERDAM	3
495	HALL	1
496	CANYON	13
497	117TH	189

3 The GIS sums the workers for each center as it assigns streets to each.

Bank Branch	# of Workers
2	3036
3	410

MAKING A MAP

After it's finished assigning the segments, the GIS automatically shows the entire network and highlights the selected segments.

If you've created a boundary to select the surrounding features, you can display it on your map. If you do, you might also want to highlight the features within the distance to make it easier to see the selected features. You can draw only the boundary outlines, or shade them. If you want readers to focus on the selected features, draw the outline only. If you want readers to focus on the areas themselves, shade them.

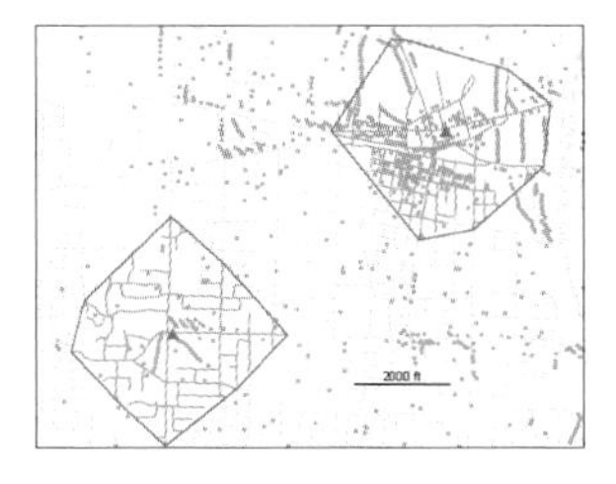

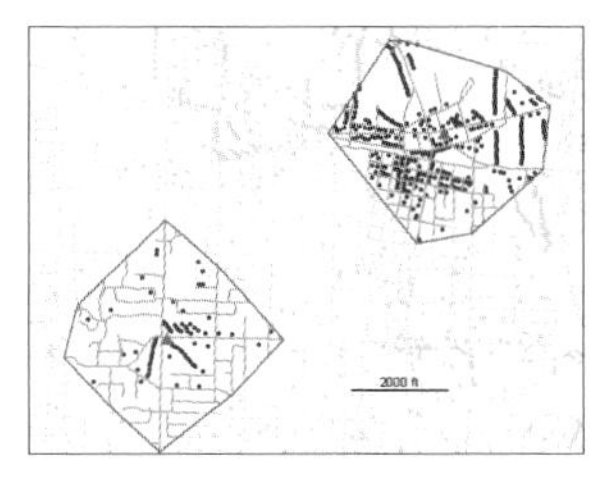

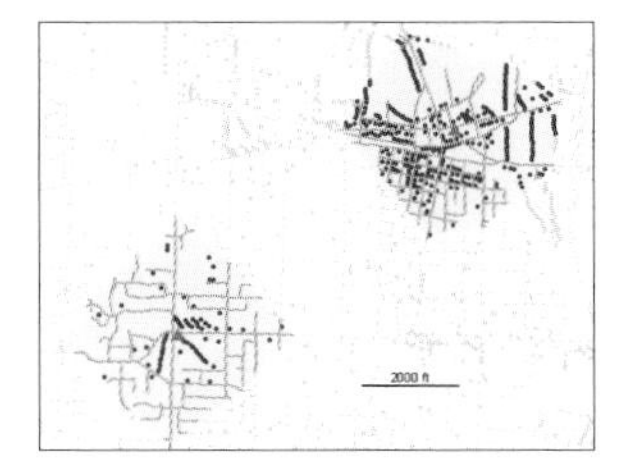

If you're showing the selected segments without the boundary, you'll want to highlight the surrounding features. It's also a good idea to place text on the map indicating what the specified distance is (for example, "0–10 miles driving distance," or "Area within a 15-minute drive").

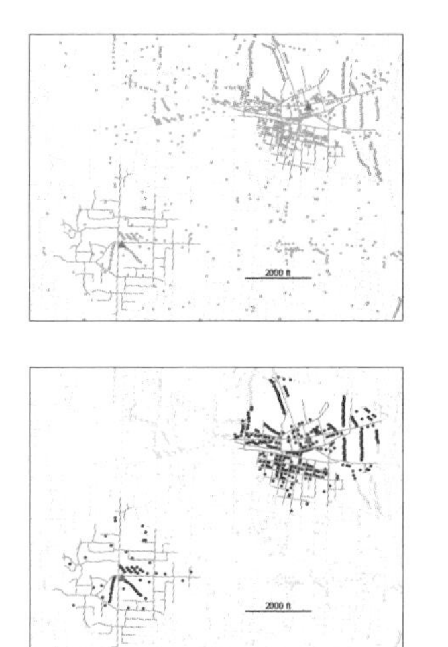

Display the center using a symbol that is easily distinguished from surrounding locations. Centers are usually shown larger and using a different symbol shape and color than the surrounding locations. You'll also want to label the centers.

Calculating cost distance over a surface lets you find out what's nearby when traveling overland. With this method, the GIS creates a raster layer in which the value of each cell is the total travel cost from the nearest source cell.

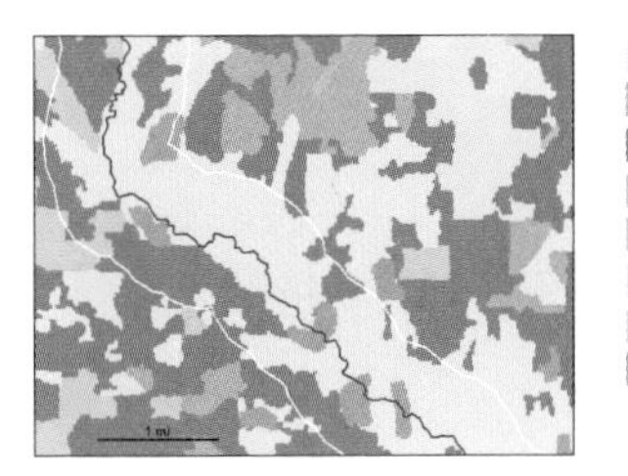

Buffer around a stream, with land cover

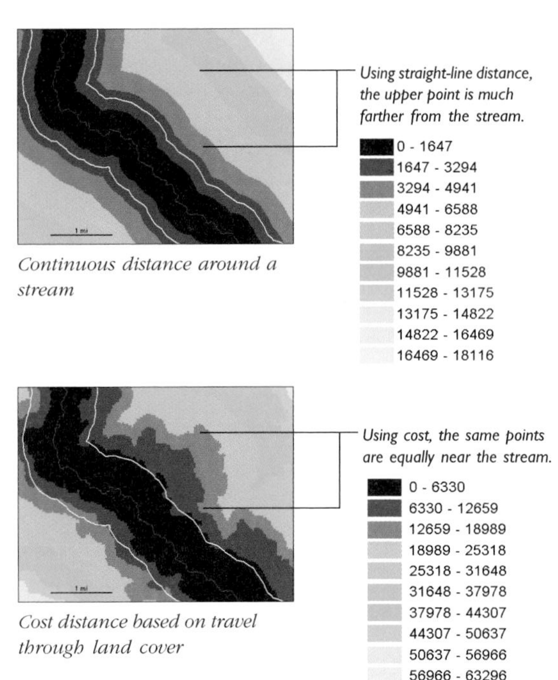

Continuous distance around a stream

Cost distance based on travel through land cover

Calculating cost over a surface also shows you the rate of change; that is, where the costs increase rapidly, and where they increase gradually. You may simply need to identify the patterns of increasing cost spreading out from the source features. But you can also find what or how much of something is within a specific cost range, or assign cost to individual features near the source.

SPECIFYING THE COST

Cost can include time; money, such as construction cost per square foot; or some other cost, such as effort expended. For example, deer might find it easier to move through open forest than thick underbrush—thus the "cost" of traveling through forest is lower.

To calculate cost over a surface, you specify the layer containing the source features and a second layer containing the cost value of each cell.

Creating the cost layer

You can create a cost layer based on a single factor or on several factors.

To create a cost layer based on a single factor, you reclassify an existing layer based on an attribute value. For example, if you know the cost per foot of building a road based on the land cover—say, 50 cents per foot through meadow, 75 cents through shrub, $1.25 through forest, and so on—you'd reclassify the land cover using these values to create the cost layer.

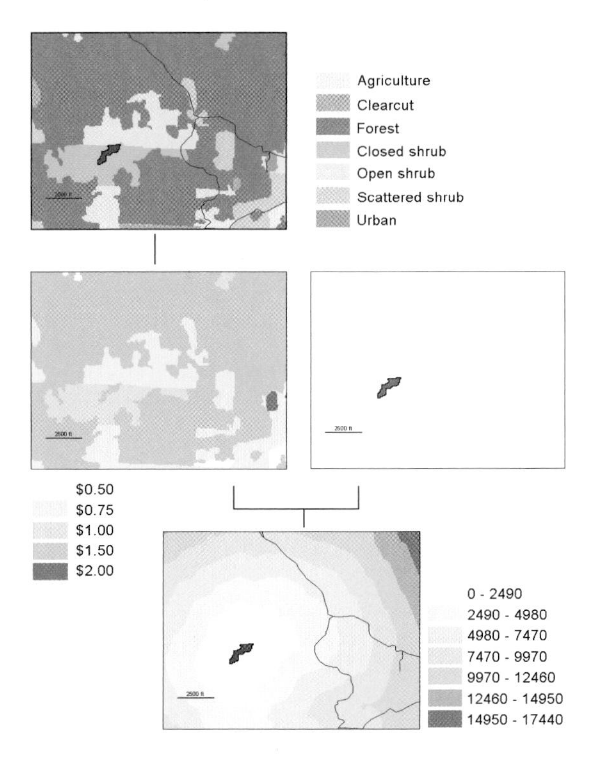

To create a cost layer based on several factors, you combine all the input layers. For example, suppose the cost for hauling timber is based on slope steepness, the type of vegetation surrounding the timber stand, and whether streams must be crossed. You'd first create three input layers by reclassifying the slope, vegetation, and streams layers using relative values, on the same scale. The slope layer would have values of 1 through 10 for least to most steep; the vegetation layer, values of 1 through 10 for various vegetation types, representing the difficulty of hauling timber through each—1 for meadow, 2 for open shrub, and so on; and the streams layer, values of 0 for no stream, and 5 for stream. You'd then combine the layers to create an overall cost surface with values ranging from 2 through 20. The GIS uses the cost layer to create the cost distance surface.

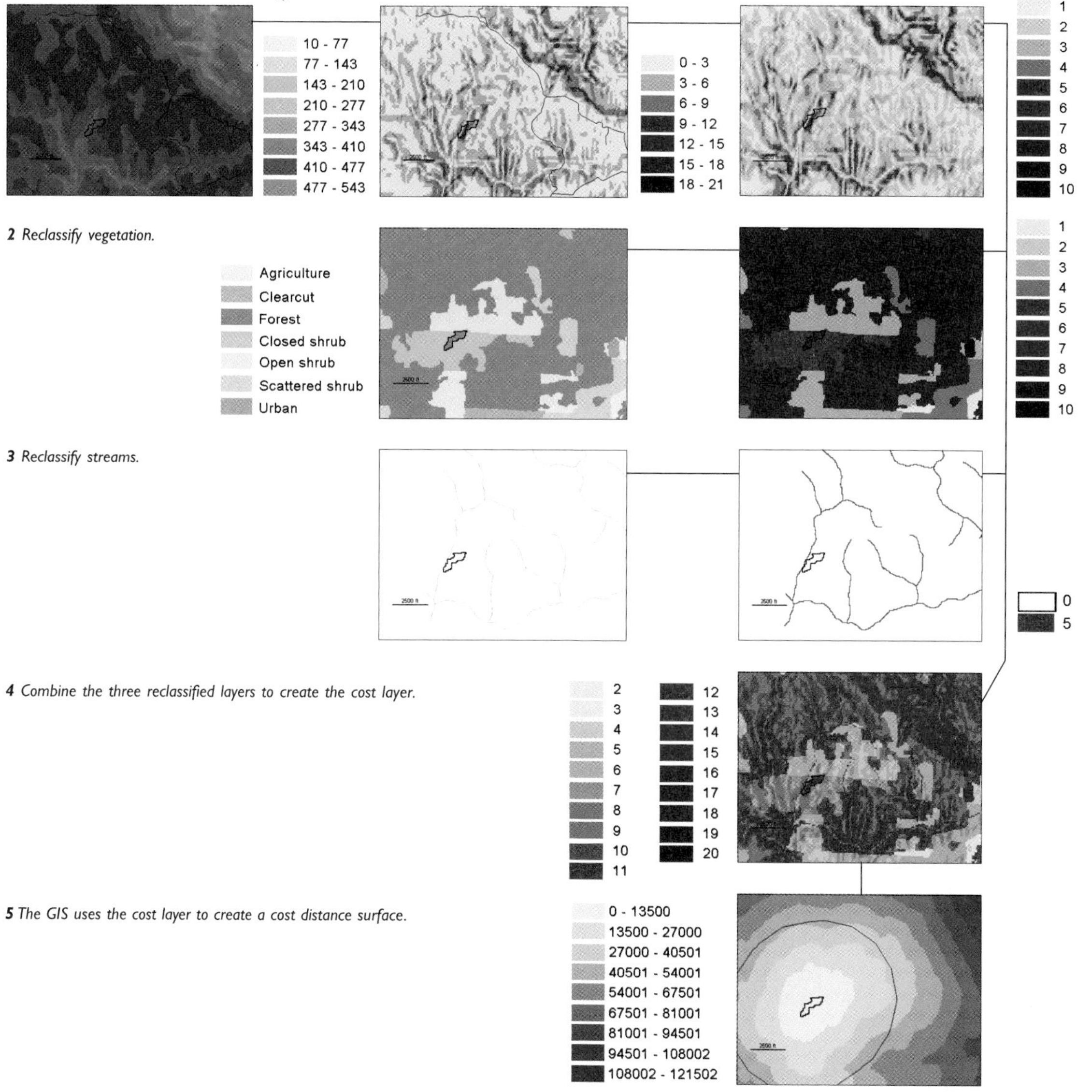

What the GIS does

Using the cost layer, the GIS totals the cost as it crosses each cell from the source, assigning a cumulative cost to each cell in a new layer it creates. Each cell is given a cost, using its size units. If the cell size is measured in meters and travel time is in seconds, the cost for a cell will be measured in seconds per meter. The GIS calculates cost from cell center to cell center; therefore, the cost to travel from one cell to the next is the sum of the cost of each multiplied by half the cell size. For example, if the cells are 50 feet on a side, one cell has a cost of one second per foot, and the adjacent cell has a cost of 10 seconds per foot, the time to travel between the cells would be calculated as:

```
(1 second/foot * (50 feet / 2)) +
(10 seconds/foot * (50 feet / 2)) =
275 seconds
```

If the travel is diagonal, the distance is slightly longer (1.4 times the width or height), so the cost is increased accordingly by multiplying the cell size by 1.4. In the example, 50 feet becomes 70 feet, and the time to travel between the cells is 385 seconds.

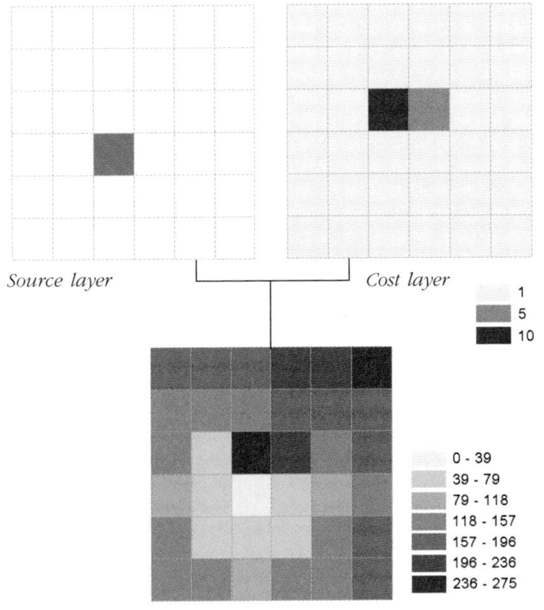

Resulting cost distance layer

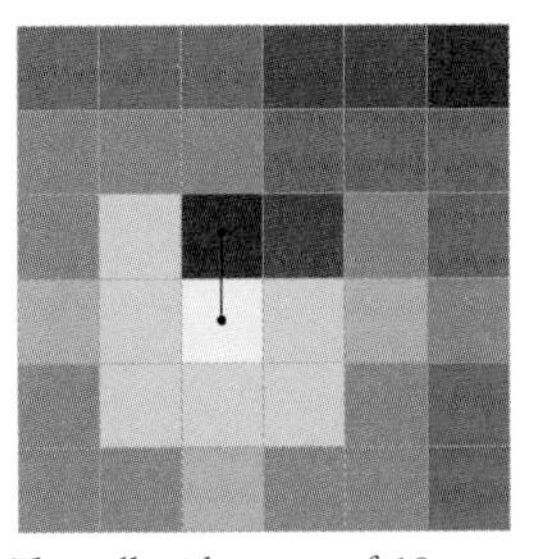

*The cell with a cost of 10 gets a cost distance value of 275 ((1 * 25) + (10 * 25)).*

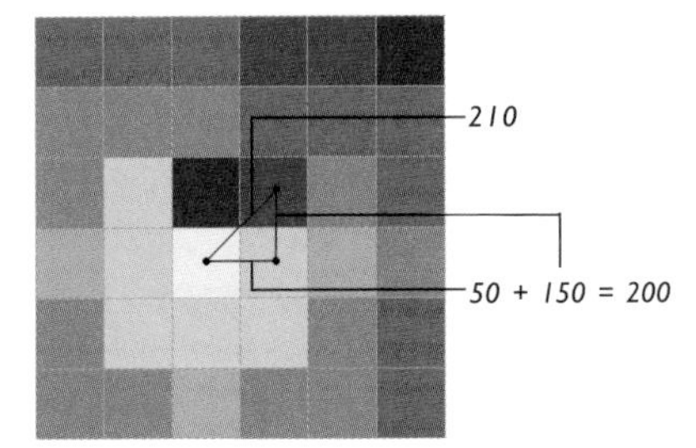

Cells are assigned the lowest cost from the source. In this example, traveling diagonally, while a shorter distance, would result in the higher cost.

The calculated value is assigned to the entire area covered by the cell. So, the larger the cell, the more approximate the value for locations farther from the cell center. Making the cell size smaller, while increasing the precision of the map, requires more processing time and storage space for the resulting raster layer.

MODIFYING THE COST DISTANCE

You can modify the cost distance surface by specifying a maximum cost or using barriers to specify areas that are "off limits."

Specifying a maximum cost

You can limit the area for which the GIS calculates cost distance values by specifying a maximum cost. The GIS stops calculating cost distance when all cells within the specified cost have been assigned a value. Any remaining cells are not assigned a value on the output layer. If you don't specify a maximum cost, the GIS calculates a value for all cells in the study area.

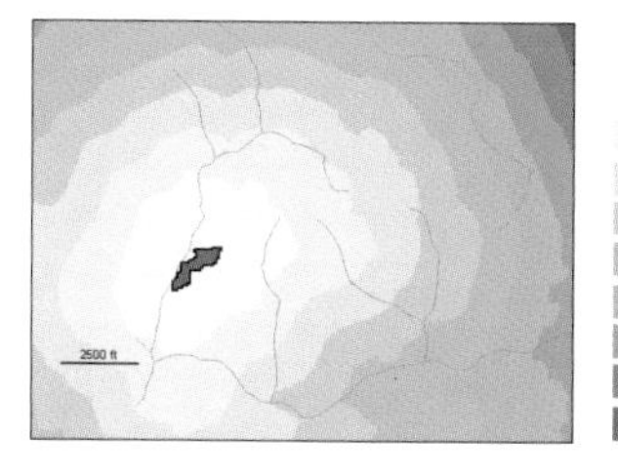

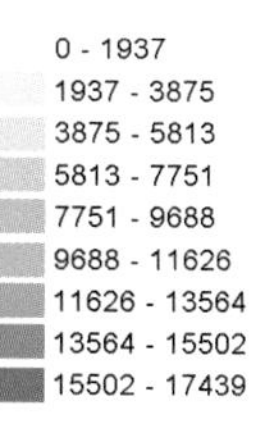

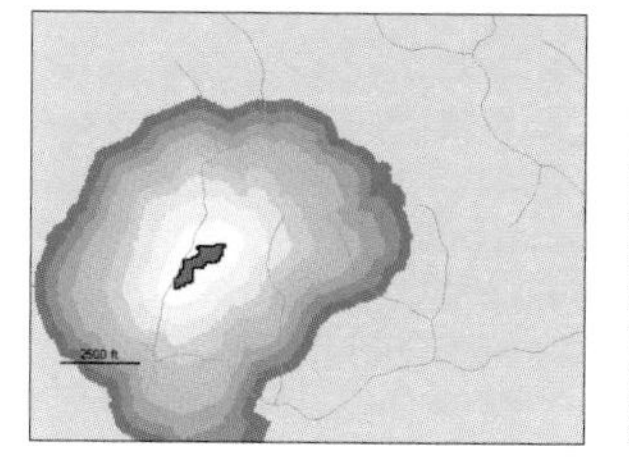

Area with cost less than $5,000

Using barriers

You can also block the assignment of cost values. For example, you might want to specify that certain areas are off limits to travel. To do this, you create a "mask" layer. Assign any "off limits" cells in the mask a value that ensures they won't be included (usually either a very high value or no value at all). All other cells should have a valid value (usually either 1 or 0). The GIS does not include the "off limits" cells when calculating cumulative cost. For example, if calculating the cost—based on land cover—to nearby forested areas, you might create a mask of logged lands over which travel is prohibited.

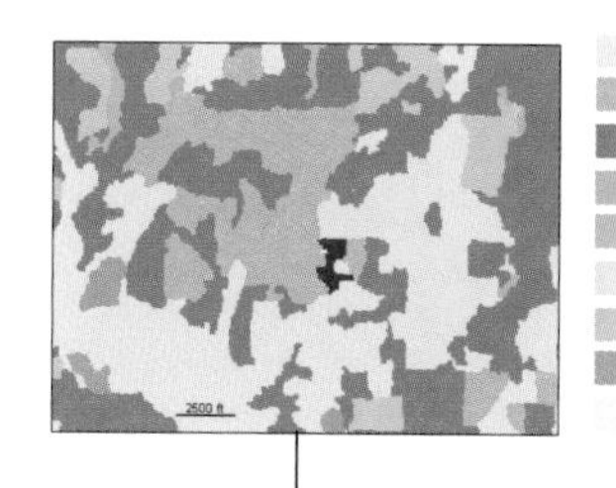

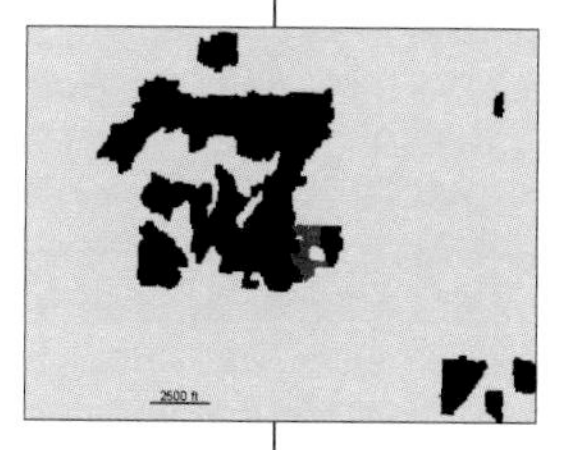

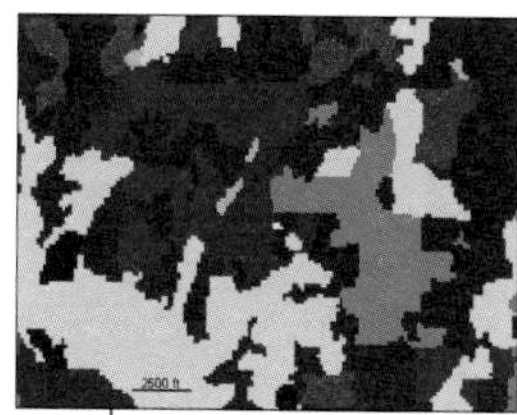

Darker reds indicate higher cost.

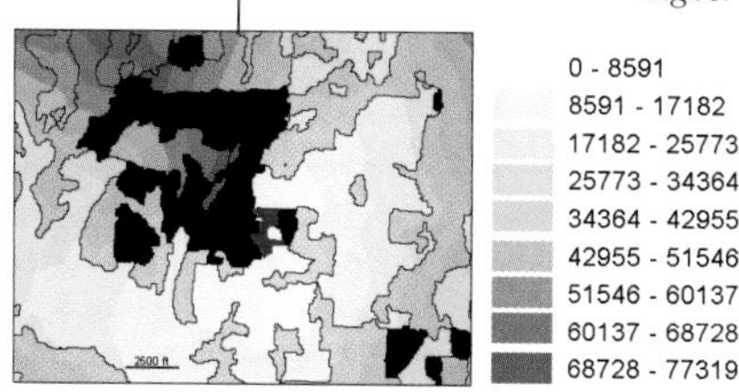

The GIS takes into account the additional travel around masked areas when calculating cost distance.

GETTING THE INFORMATION

Once the GIS has created the cost distance layer, you can either identify the area within a specific distance of the source features or summarize how much of something is within the distance.

Identifying the area within the cost

The output layer is a surface of increasing cost values radiating outward from the source. If you need to know what's within a specific cost—say, $5,000—you select cells with a value of less than or equal to $5,000 to create a new layer. To create multiple distance buffers, you can simply reclassify the grid into cost ranges, for example 0–$5,000, $5,000–$10,000, $10,000–$15,000, and $15,000–$20,000.

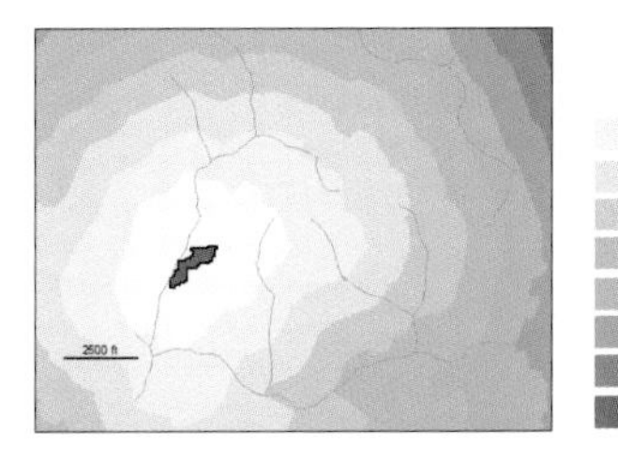

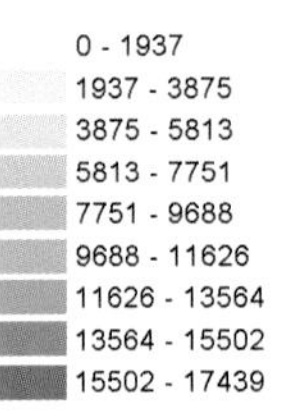

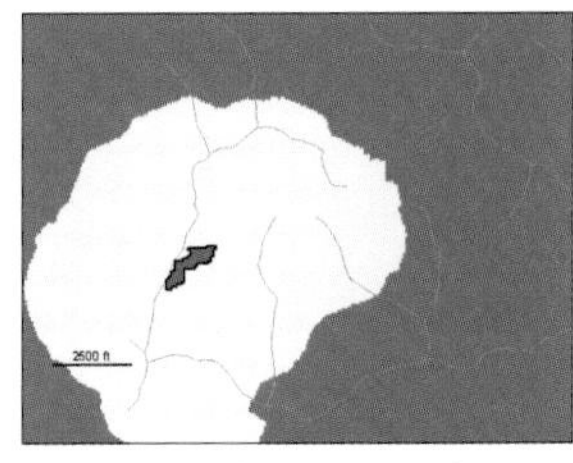

Area within $5,000 cost distance

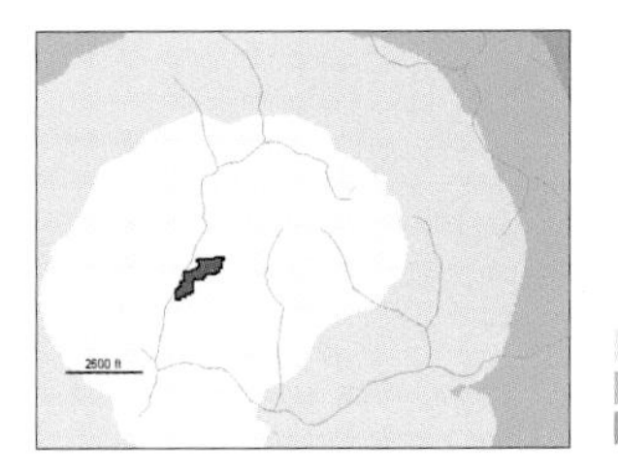

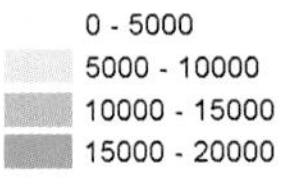

Summarizing what's within the distance

You can also summarize what's within a given cost distance, or assign the cost to individual features, the same as you would with straight-line distance over a surface (see the earlier section 'Distance over a surface'). You reclassify the surface into one or more ranges, then combine it with the layer containing the surrounding features. For example, you might want to calculate the amount of forest and other land cover types within a $5,000 haul cost of a timber stand.

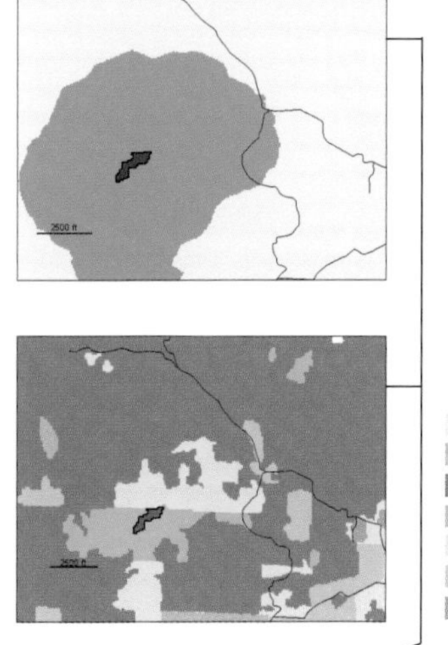

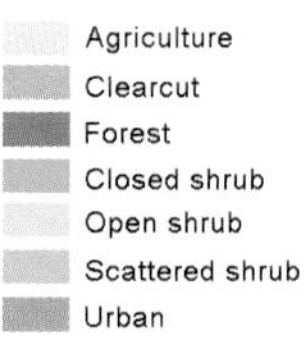

Land Cover	Acres
Forest	1132
Closed shrub	342
Agriculture	0
Clearcut	34
Scattered shrub	19
Open shrub	498
Urban	0

MAKING A MAP

If you're mapping discrete features with the cost distance surface, you can show them on top of the distance grid. The distance grid is displayed using graduated colors.

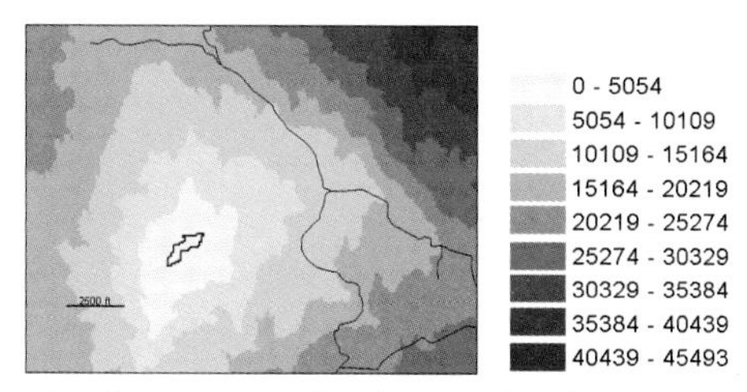

Cost distance around timber stand, with roads

If you specify more than six or seven ranges, you can use two or three hues to help distinguish the ranges. You'll also want to show the source features in a contrasting color.

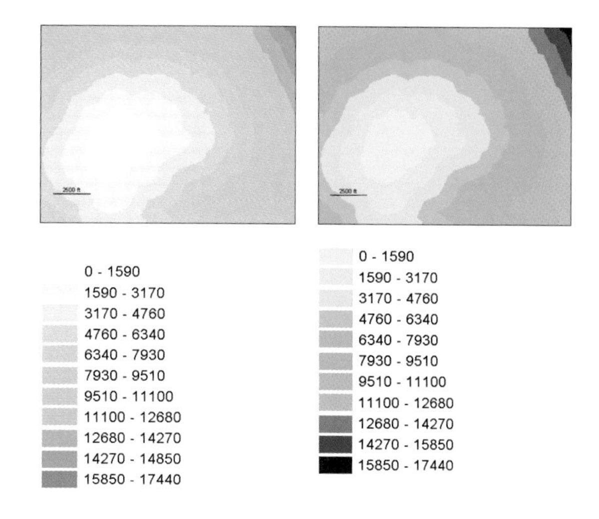

7 Mapping change

GIS lets you map where things move, or the changing conditions in a place over time. Knowing what's changed can help you understand how things behave over time, anticipate future conditions, or evaluate the results of an action or policy.

In this chapter:

- *Why map change?*
- *Defining your analysis*
- *Three ways of mapping change*
- *Creating a time series*
- *Creating a tracking map*
- *Measuring and mapping change*

People map what's changed to anticipate future conditions, decide on a course of action, or to evaluate the results of an action or policy.

By mapping where and how things move over a period of time, you can gain insight into how they behave. For example, a meteorologist might study the paths of hurricanes to predict where and when they might occur in the future. A wildlife biologist might study the movement of bears over a 24-hour period to understand their foraging habits and know how much land must be set aside to maintain the population.

Another reason for mapping change is to anticipate future needs. For example, a police chief might study how crime patterns change from month to month to help decide where officers should be assigned. A transportation planner might look at trends in traffic flow to see where lanes will need to be added to highways and streets in the future.

By mapping conditions before and after an action or event, you can see the impact. A police chief, for example, might map the locations of narcotics arrests in a neighborhood for the six months before and after a drug sweep, to see how effective the crackdown was. A retail analyst might map the change in store sales before and after a regional ad campaign to see where the ads were most effective.

MAP GALLERY

Since just after World War II, the Emaralda Marsh, home to birds and other wildlife, has slowly been shrinking due to development around its edges. To create a restoration strategy for the marsh, located northwest of Orlando, Florida, the St. Johns River Water Management District mapped vegetation for four periods from 1941 to 1987. The maps help biologists compare historical vegetation patterns and understand how current conditions developed. The information will help the district restore the wetland to create more waterfowl habitat, provide recreation, and improve water quality.

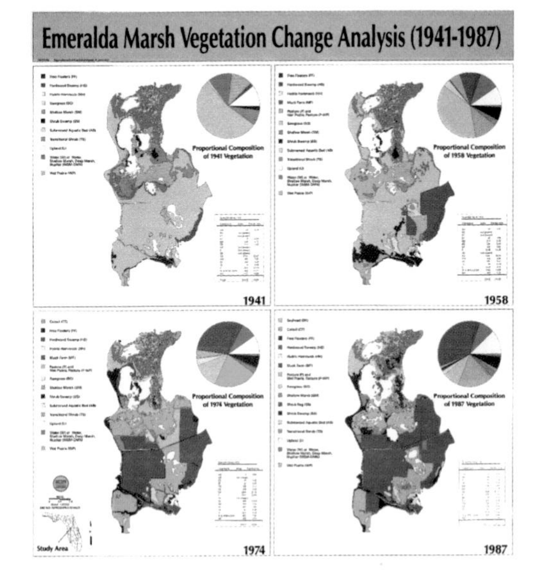

While air quality has improved in recent years, pollution continues to be a major problem in parts of California. To help policy makers develop more effective regulations, the Air Resources Board created a series of statewide maps showing the trend in car and truck registrations, miles traveled, and exhaust emissions, by county, from 1975 to 1995. The maps show that while the number of vehicles on the road has increased across the state, some emissions have actually decreased. Others have increased in some areas, indicating where regulations should be strengthened.

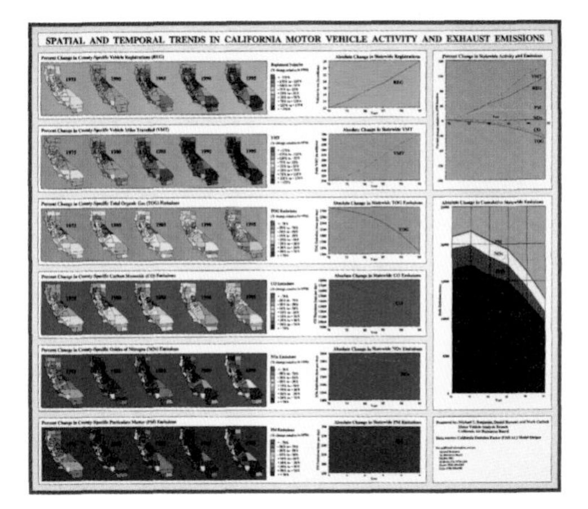

The South Florida Water Management District is responsible for providing flood protection—including protection against flooding from hurricanes—to the nearly six million people living within its jurisdiction. As part of an effort to provide information to the public on how the District plans for and responds to hurricanes, analysts created a set of maps showing how many hurricanes have hit Florida during storm season (June–November) since 1886.

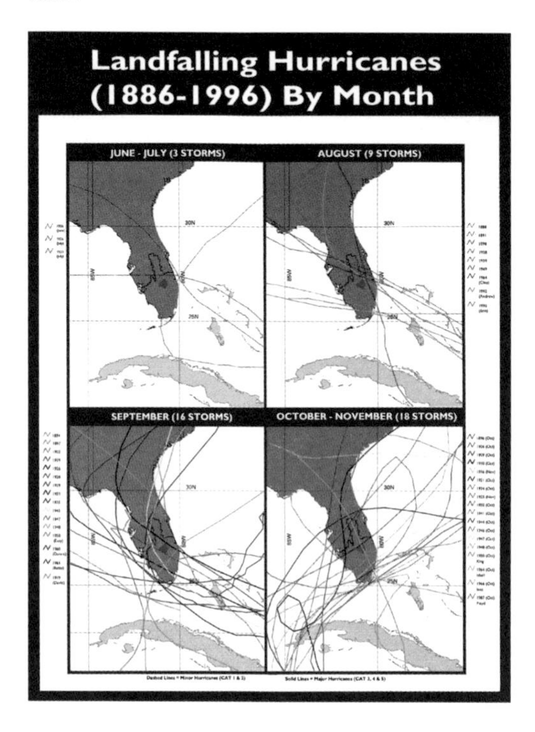

The Taylor Valley of Antarctica is a unique area of glaciers, steep slopes, broad valley floors, and high plateaus. As part of a study on the ecology of this region, researchers at Colorado State University created a series of maps showing the hours of direct sunshine covering the surface over a 24-hour period, on the first day of each month. The scientists use the maps to help understand the effect of sunlight on the climate, glaciers, soils, lakes and streams, and biology of the area.

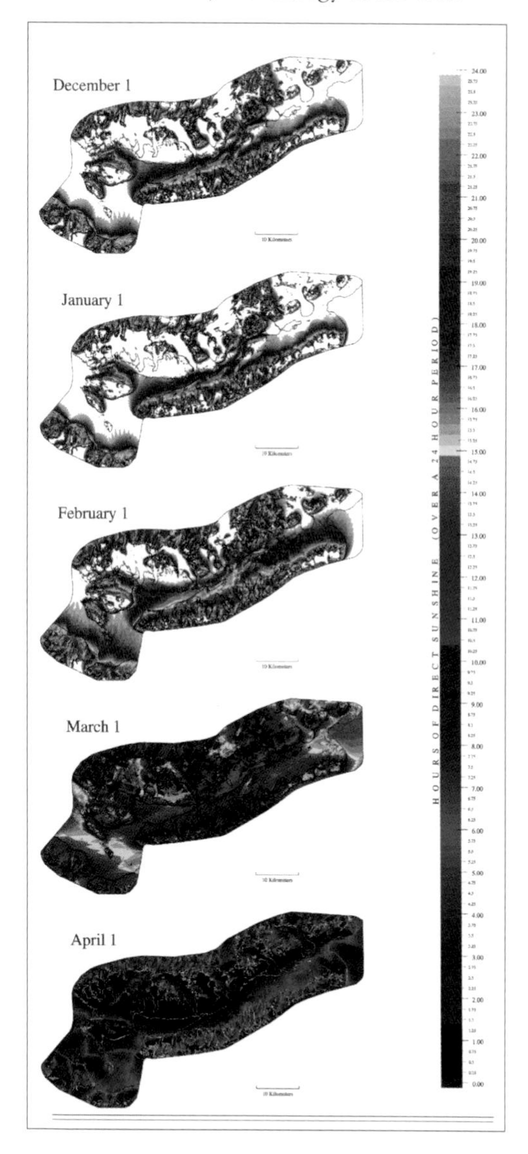

You can map change by showing the location and condition of features at each date, or you can calculate and map the difference in a value for each feature between two or more dates.

Knowing the type of change and the type of features you're dealing with, how you're measuring time, and the type of information you need from the analysis will help you decide how to map change.

TYPES OF CHANGE

Geographic features can change location, or change in magnitude or character.

Change in location

Mapping change in location helps you see how features behave so you can predict where they'll move. For example, you might map the paths of hurricanes to see whether the patterns change from month to month. Or, you could map where peregrine falcons fly during the course of a migrating season, to see their range.

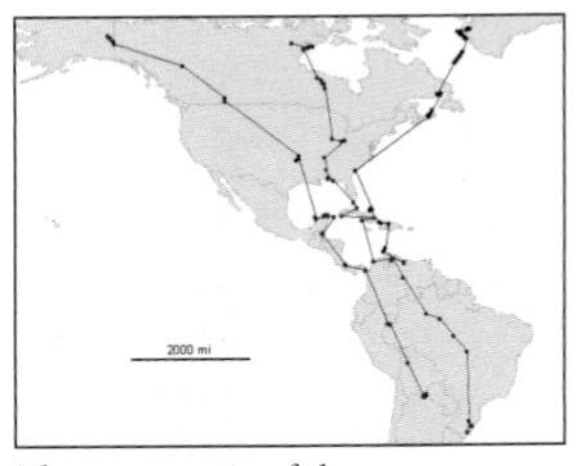

Three peregrine falcons were tracked as they migrated south.

Change in character or magnitude

Mapping change in character or magnitude shows you how conditions in a given place have changed. The change can be in the type of feature in a place—for example, the different categories of land cover in a watershed now, compared to 20 years ago.

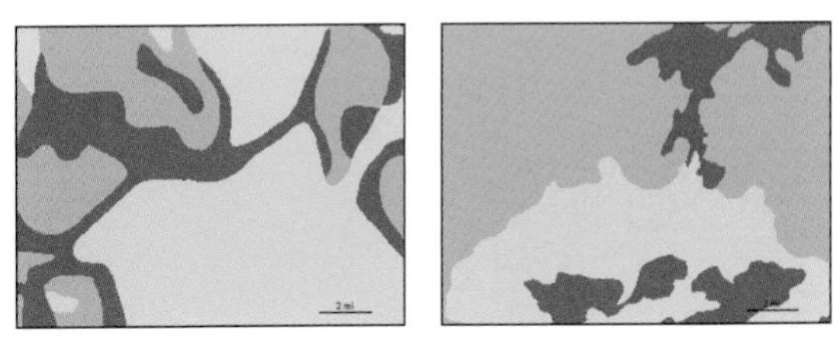

Land cover in 1914 (left) and 1988

Or, the change can be in a quantity associated with each feature—for example, the amount population has increased or decreased in each county over the past 20 years, or the change in carbon monoxide readings at a monitoring station from season to season.

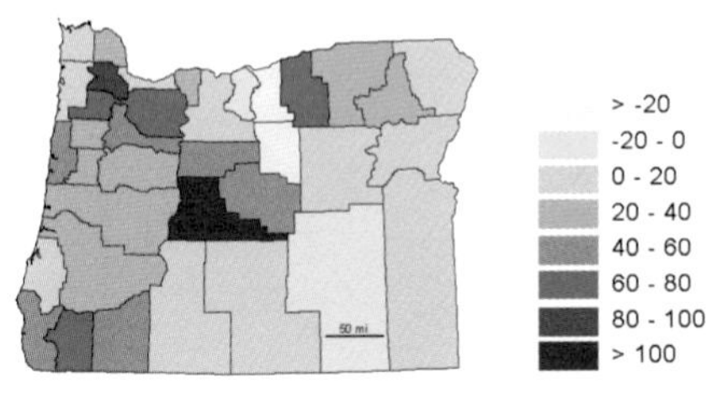

Percentage population change per county, 1970–1990

Change in location and character are not mutually exclusive. Something may change location and magnitude at the same time, such as a hurricane in which the wind speeds vary as the hurricane moves over water and land.

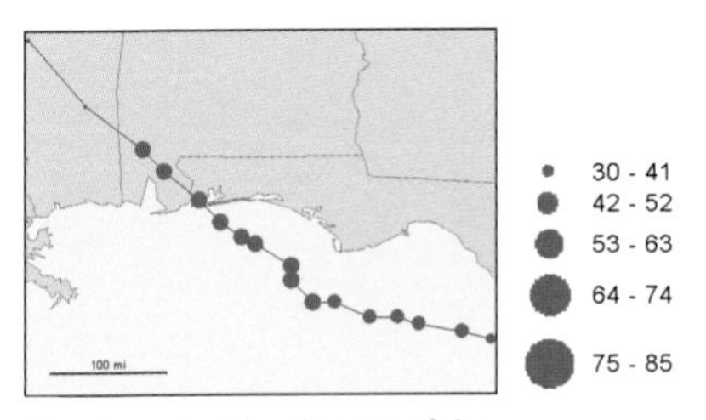

Hurricane track—the size of the dots indicates the relative wind speed, in miles per hour.

THE GEOGRAPHIC FEATURES

The type of features you're mapping help you determine the best method to map change.

Features that move

You can map discrete features that physically move, or events that represent geographic phenomena that change location.

Discrete features

Discrete features can be tracked as they move through space. They might be individual features you can map paths for, such as a hurricane, a vehicle, or an animal; linear features, such as a stream channel that changes position; or an area feature, such as the boundary of a fire you can delineate at any given time. Area features often represent an edge that expands or contracts, such as an oil spill, a wildfire, or the boundary of the developed area surrounding a city.

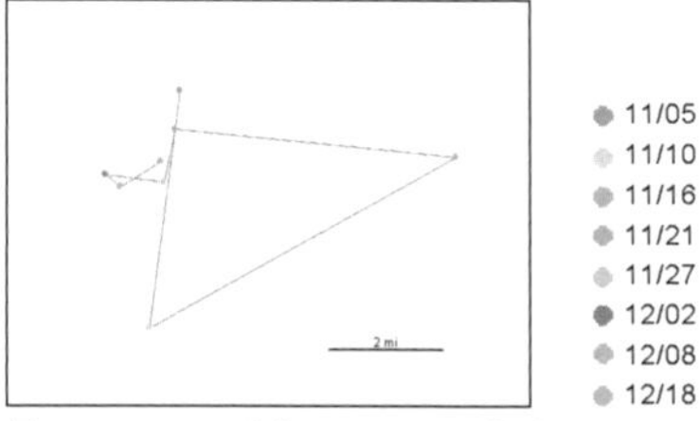

This peregrine falcon was tracked over several weeks to show the extent of its home range.

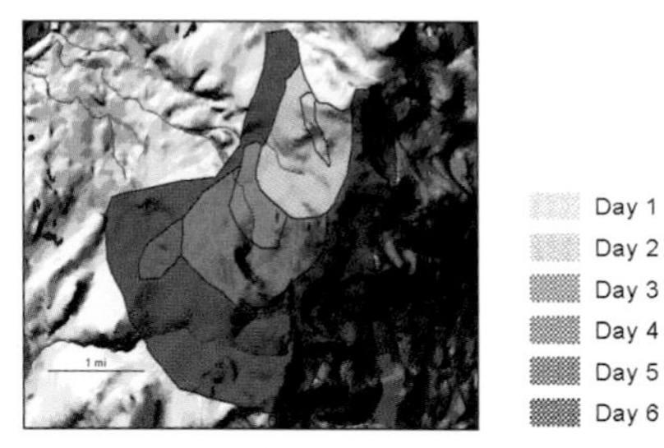

This map shows the spread of a wildfire over six days.

Events

Events, such as crimes or earthquakes, represent geographic phenomena that occur at different locations. While each individual event occurs at a specific location at a specific instant, the set of events can be tracked and mapped to show the movement of the phenomena over a period of time. For example, by mapping calls to 911 reporting drug-related activity, you can get a sense of where drug dealing has moved over several months.

Features that change in character or magnitude

You can map change in character or magnitude for discrete features, data summarized by area, continuous categories, or continuous numeric values.

Discrete features

These are features that change in character or in the quantity of an attribute associated with them. Examples include: stores for which sales change from month to month, parcels for which the land use changes over a 10-year period, or streets for which the traffic volume changes over a 24-hour period.

Data summarized by area

These are totals, percentages, or other quantities associated with features within defined areas, such as the population in each county for each year, or the number of calls to 911 in each neighborhood for each month.

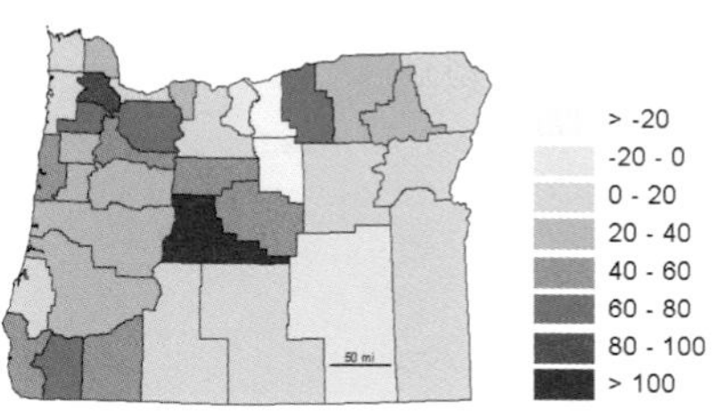

Percentage change in population, 1970 to 1990

Continuous categories

Continuous categories show the type of features in a place, such as each land cover type. They can be represented by boundaries or as a surface.

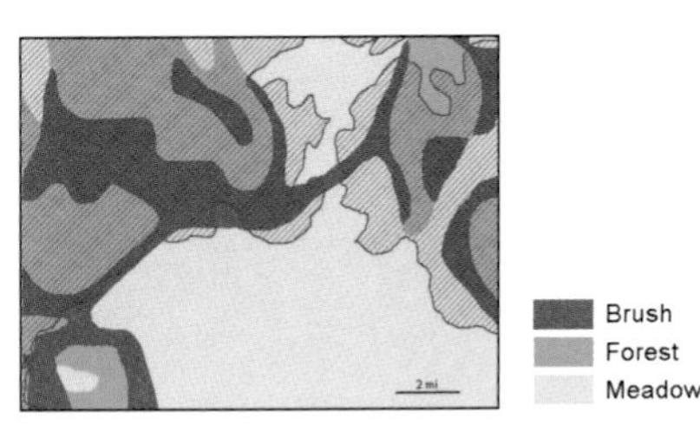

1914 land cover with 1988 forest boundary (hatched area)

Continuous values

These are quantities that are continuous, such as air pollution levels. At any location, there is a measure of the values. This data is often monitored at fixed points—such as air quality monitoring stations—and interpolated to create a surface.

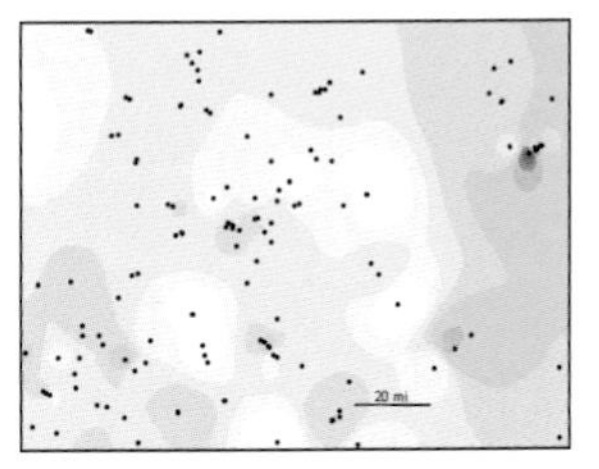

Surface of maximum eight-hour carbon monoxide (CO) readings over the course of a year, with monitoring station locations. The lightest areas had the lowest levels of CO.

MEASURING TIME

The time pattern you're mapping, and how you partition the time period you're dealing with, will affect the geographic patterns you see on your map.

The time pattern

You can map three types of time patterns:

- A trend—change between two (or more) dates or times
- Before and after—conditions preceding and following an event
- A cycle—change over a recurring period of time, such as a day, month, or year

Trends indicate whether something is increasing or decreasing, or the direction of a feature's movement; cycles show recurring patterns that reveal information about the behavior of the features you're mapping. Mapping conditions before and after an event or action will let you see the impact.

You can look at a single set of data with each of these time patterns. For example, if you have sulfur dioxide levels over 24 hours for 10 years, you could map the average annual level each year to see where pollution is increasing or decreasing. Or, you could summarize the values by six-hour period over the course of a single year to see the daily cycle of pollution levels. Or you could map average levels for the year before and after a specific date, such as the date a new regulation went into effect, to see the impact of the regulation.

You can also combine these patterns. For example, you could show how the daily cycle of pollution has changed over the course of several years.

Partitioning time

You can display feature locations or characteristics at two or more times or dates, or you can summarize feature attributes over a period of time or several time periods. You'll also need to decide how many times or dates to map, and the interval between them.

Using a snapshot or a summary

Snapshots show the condition at any given moment, and are used to map phenomena that are continuous in time, such as population, land cover, or air quality. At any instant, there is some value for the phenomena.

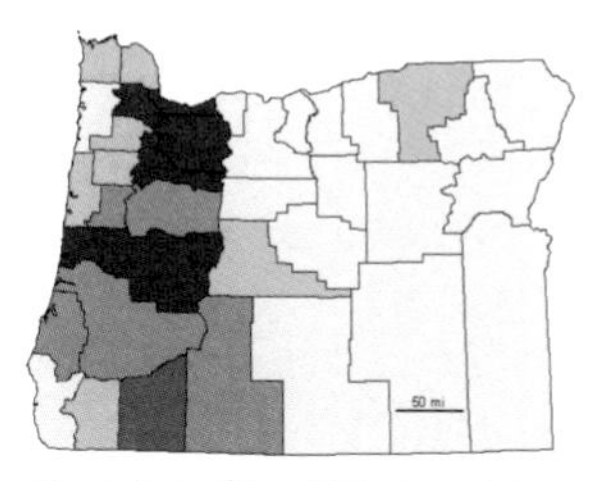

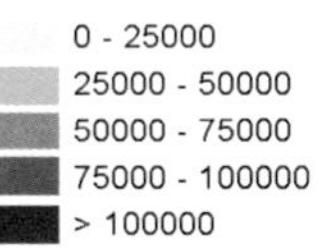

Snapshot of the 1970 population by county, as of the date of the census

Summarizing is used for mapping discrete events in a particular place that are not continuous in time—that is, at any instant, an event either is or isn't occurring. For example, you'd map calls to 911 occurring over a month, or the location of earthquakes over several years (mapping the earthquakes occurring at any given instant might make for a very sparse map). You can also summarize a value of a continuous phenomenon over a given time period. For example, you could summarize daily temperatures for a city into monthly averages.

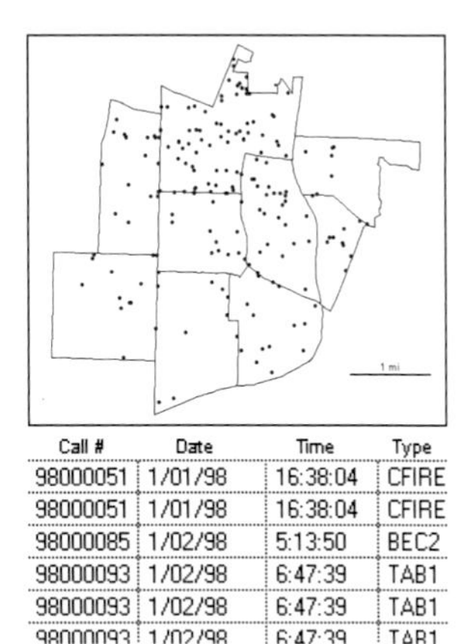

Call #	Date	Time	Type
98000051	1/01/98	16:38:04	CFIRE
98000051	1/01/98	16:38:04	CFIRE
98000085	1/02/98	5:13:50	BEC2
98000093	1/02/98	6:47:39	TAB1
98000093	1/02/98	6:47:39	TAB1
98000093	1/02/98	6:47:39	TAB1

Calls to 911 in several neighborhoods, for the period from January 1 to January 31

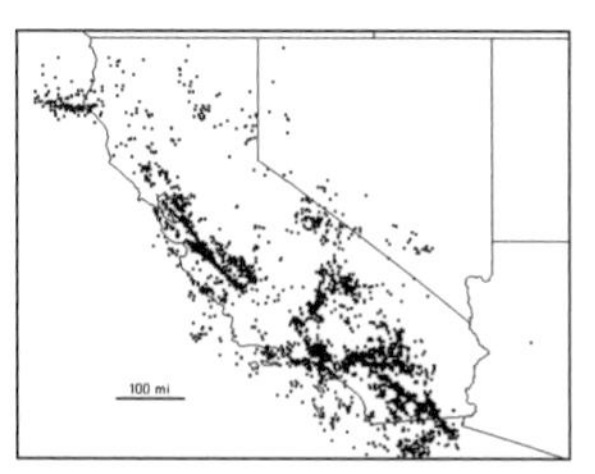

Earthquakes, 1970 to 1974

When summarizing, you can either count the occurrences of events (such as the number of auto thefts), sum a value (such as the total monetary value of the thefts), or calculate other statistics (such as the average value of each theft).

You can also summarize and map discrete events within a set of boundaries—for example, the total number of calls to 911 in each neighborhood each month. You can also map features by summarizing what occurs near them. An example is the number of lightning strikes near each power pole during a storm.

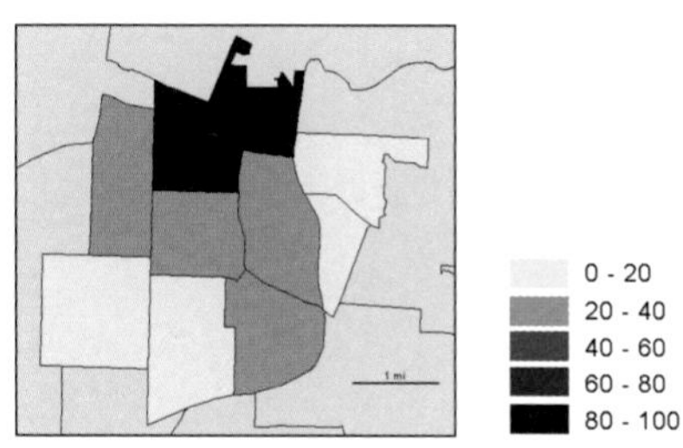

Neighborhoods shaded by number of calls to 911, from January 1 to January 31

Choosing the number of dates and the interval

If you're mapping trends, you need to determine the interval, the number of dates, and the duration, or total period. The duration divided by the number of dates yields the interval. Conversely, the duration divided by the interval yields the number of dates.

The interval is already determined if the data has been collected on only a few dates. However, if you have a range of dates, you can choose the interval. For example, if you have annual population by county over a 20-year period, you could show the population every year, every five years, or every 10 years. It's easier for map readers to understand the patterns if you use a regular interval. The interval should be long enough to show change between maps, but not leave information out. For example, the population change from year to year may not be very dramatic, but from decade to decade there may be noticeable changes. Charting your data lets you see the distribution of values and can help you choose an appropriate interval. If the change is rapid, use a shorter interval; if the change is gradual, use a larger interval.

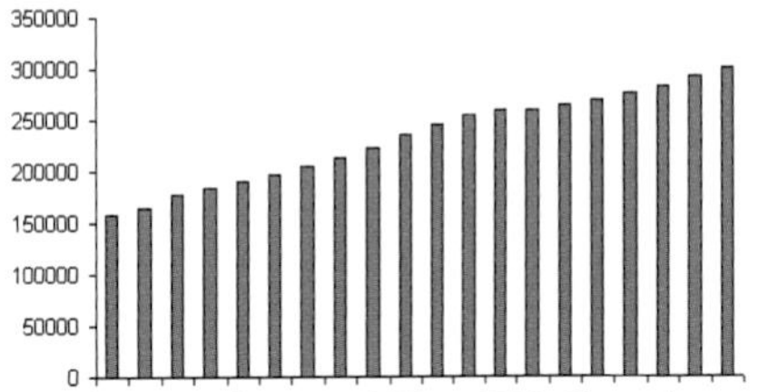

Population each year, 1970 to 1989. Since the increase is gradual, using five- or 10-year intervals would show the change.

The number of dates you use depends on the consistency of the change. If the change tends to be slow and steady, a few dates, widely spaced, may accurately capture the change. However, with fewer dates and a wider interval, you may miss changes that occur between the dates. You also won't know during which period most of the change occurred. For example, mapping population in 1970 and 1990 shows substantial change between the dates. But if you map population in 1970, 1980, and 1990 you can see that most of the change occurred between 1970 and 1980.

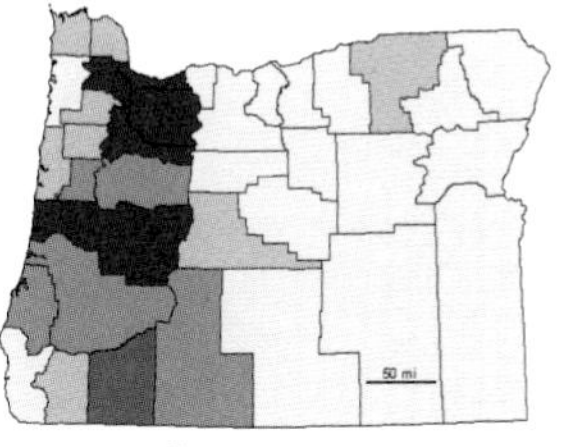
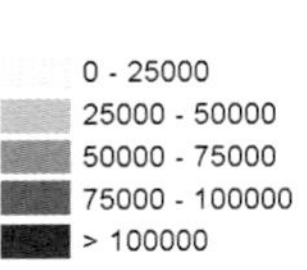

1970 population

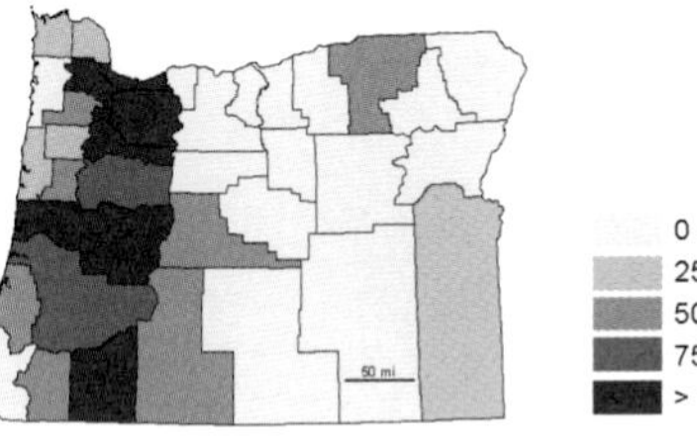
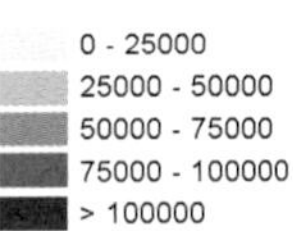

1980 population

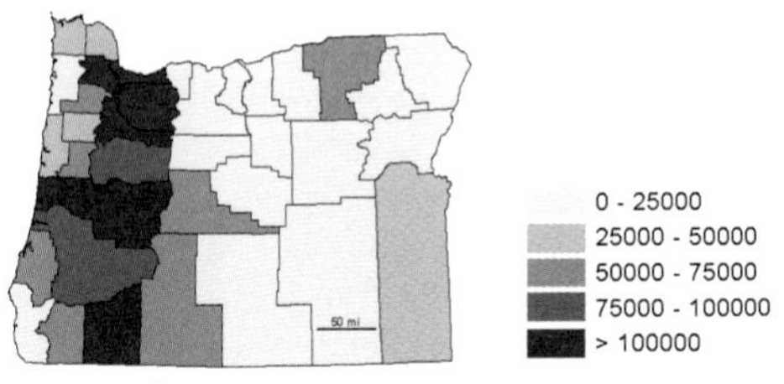

1990 population

If you're mapping cycles, you can either show a snapshot or summarize over a period, depending on whether you're mapping discrete events or continuous data.

For discrete events, you summarize the events rather than use a snapshot. For example, rather than mapping calls to 911 that occurred exactly at 9 A.M., 3 P.M., 9 P.M., and 3 A.M., you'd assign a code to each call to indicate whether it occurred in the morning (6 A.M. to noon), the afternoon (noon to 6 P.M.), and so on; you'd then map the calls based on the period they fall within. This gives you a larger sample, and can make the patterns easier to see.

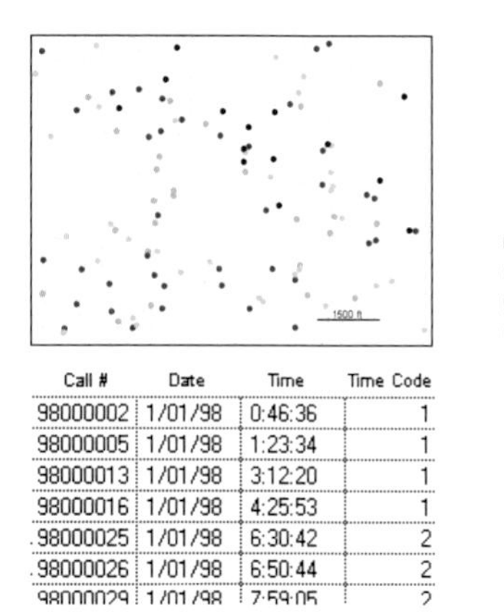

Call #	Date	Time	Time Code
98000002	1/01/98	0:46:36	1
98000005	1/01/98	1:23:34	1
98000013	1/01/98	3:12:20	1
98000016	1/01/98	4:25:53	1
98000025	1/01/98	6:30:42	2
98000026	1/01/98	6:50:44	2
98000029	1/01/98	7:59:05	2

For continuous data, you could use a snapshot, for example, showing the ozone reading at 9 A.M., 3 P.M., 9 P.M., and 3 A.M. Alternatively, you could summarize by period, for example, show the average ozone value between 6 A.M. and noon, between noon and 6 P.M., and so on. Summarizing smooths high and low values, making it easier to compare values over time.

In either case, you'll need to subdivide the cycle. For example, you could map a daily cycle by 24 one-hour periods, four six-hour periods, or two 12-hour periods. Or you could map an annual cycle by twelve months, or four seasons. Use the fewest divisions necessary to show change. The more variable the change, the more divisions you'll want to use. For example, if you're mapping rainfall in California, showing fours maps of rainfall by season would be enough, since the change between seasons is greater than the change month to month.

You also need to decide what the duration is—that is, how many cycles to include. You might map monthly precipitation over the past 10 years, or the past 30 years. You'll want a long enough duration to minimize any anomalous events. For example, if you calculated seasonal rainfall in Los Angeles over five years, and two of these happened to be El Niño years, you'd conclude that average winter rainfall was much higher than if you had calculated rainfall over a 20-year period.

If you're mapping conditions before and after a catastrophic event, such as a hurricane or fire, you'd use snapshots before and after the event. You'll want to pick dates as close to the event as possible to better assess the impact of the event itself. The longer the lapse, the more other factors may come into play.

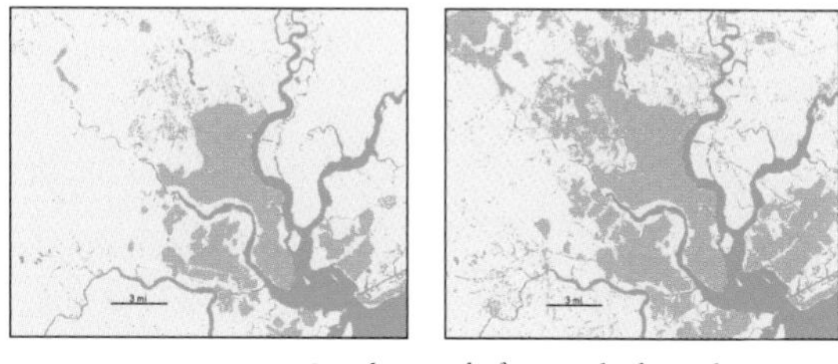

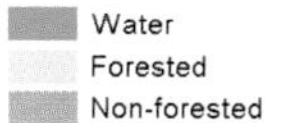

Land cover before and after a hurricane

If you're mapping an ongoing activity before and after an action or other event, you'd summarize over a period. The period should be long enough to see the results of the event or action. For example, you may be able to see within a few weeks or months after a police crackdown the change in drug dealing in a neighborhood, while the impact from a change in air pollution regulations for a state could take several years to become apparent.

THE INFORMATION YOU NEED FROM THE ANALYSIS

If you're mapping change in magnitude or character, you can measure and map how much or how fast a place changed. Calculating a change in value, rather than simply mapping the conditions at two different times, highlights the features that have changed the most or the least.

How much it changed

When calculating change in magnitude, you subtract the numeric values associated with each feature. For example, you might subtract the 1970 population for each county from the 1990 population to get the population change between 1970 and 1990. Or subtract sales for each store for 1997 from the sales in 1998 to get the change in sales between the two years.

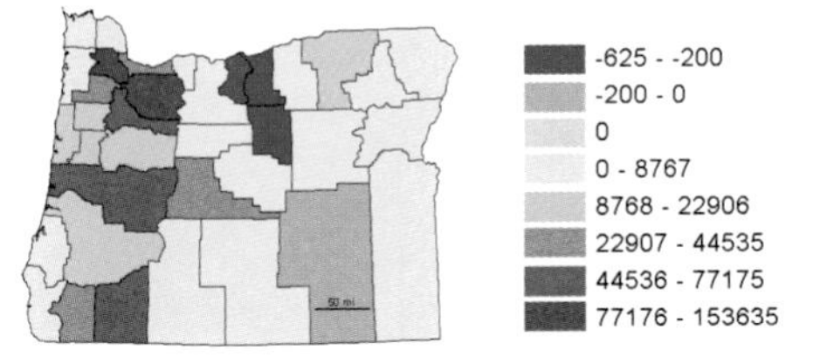

Increase (or decrease) in population, 1970 to 1990

You can also calculate the percentage change by dividing the difference by the original value and multiplying by 100. Mapping a percentage shows you which features changed the most relative to their original value, and is particularly useful when the features vary in size. For example, you'd map percentage change in population for counties to see which ones are growing fastest. Counties with a large population might have a larger increase in numbers even though the percentage increase is much smaller.

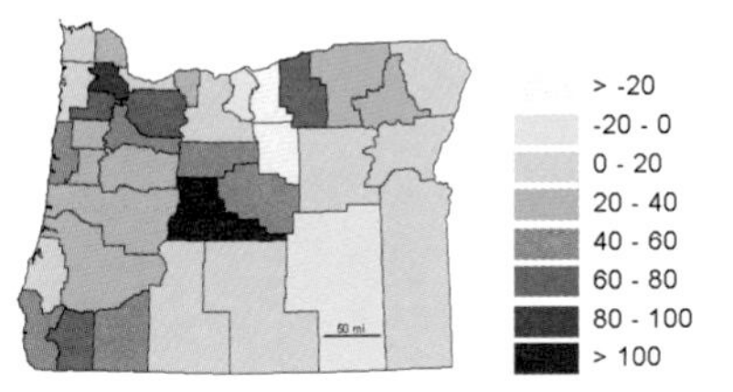

Percentage increase or decrease in population, 1970 to 1990

To measure change in type or category, you sum the land area of each category and calculate the actual or percentage difference between the dates. For example, you'd sum the acres of standing forest before and after a hurricane, and subtract the values to find out how much forest was destroyed.

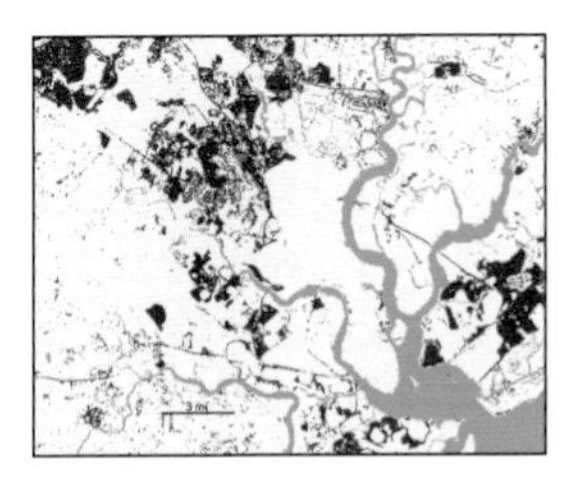

How fast it changed

You can also measure the rate of change. To do this, you divide the difference between the two dates by the number of units in the time period, to get the average change per time unit. For example, you could subtract the 1970 population in each county from the 1990 population, and divide this value by 20 to get the population change per year. This is an average over the twenty-year period and does not necessarily indicate the actual change in any given year. It can be useful, though, for comparing features. For example, you could map each county by its rate of population change to see which grew rapidly and which slowly.

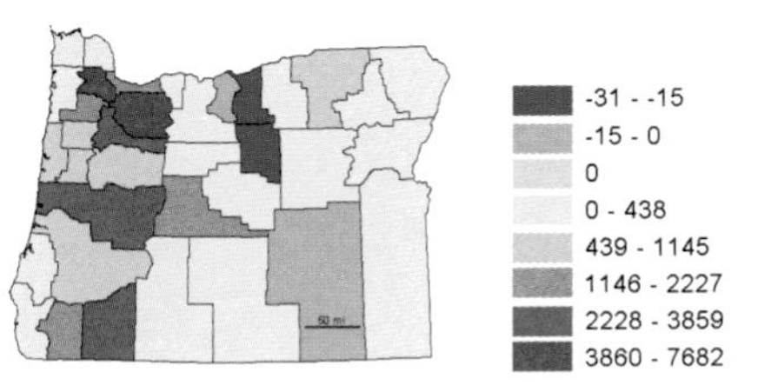

Average increase or decrease in population each year, 1970 to 1990

You can map change using a time series or a single tracking map, or you can measure and map the difference in values between two times or dates.

Creating a time series

A time series is good for showing changes in boundaries, values for discrete areas, or surfaces. You create one map for each time or date showing the location or characteristics of the features.

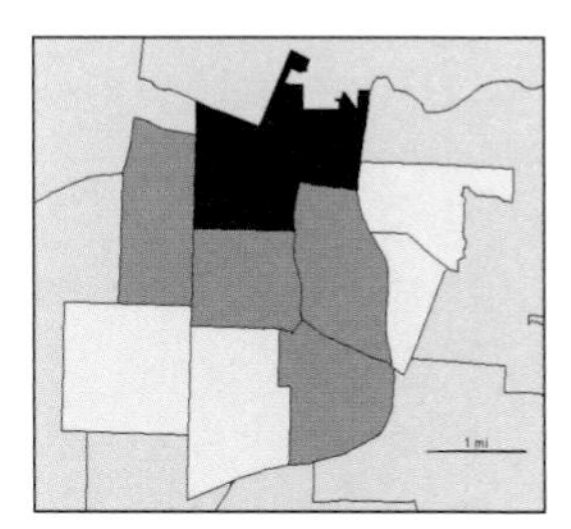

January

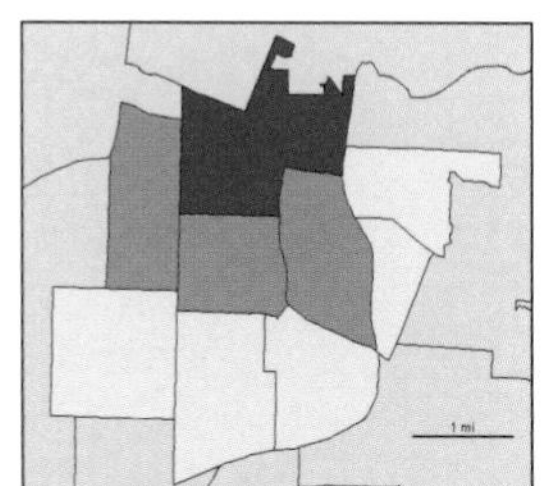

February

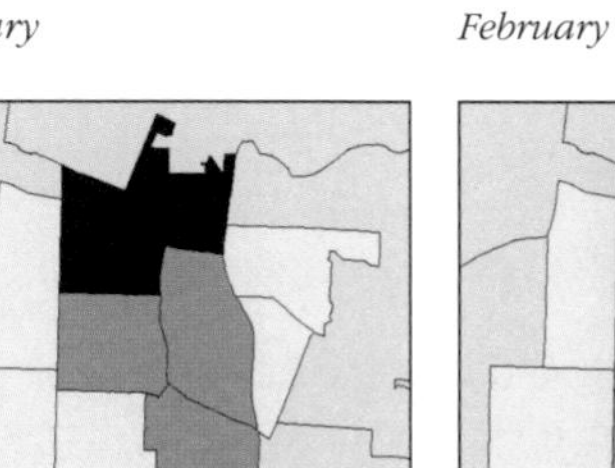

March

April

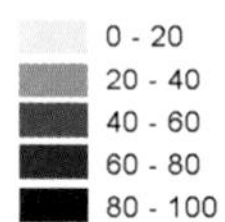

Time series of calls to 911, by neighborhood

Creating a tracking map

A tracking map is good for showing movement in discrete locations, linear features, or area boundaries. You create a single map showing the locations of the features at several dates or times.

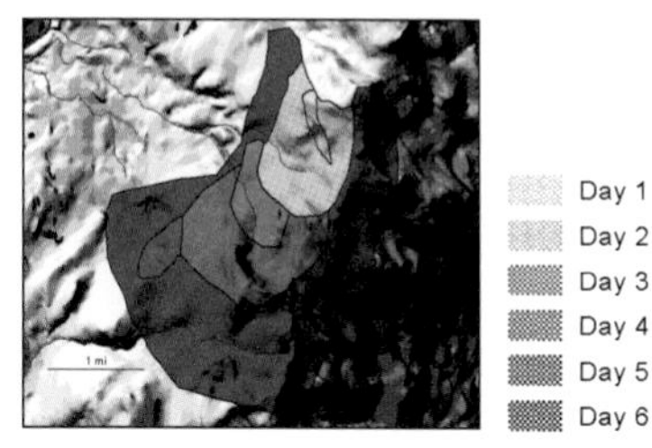

Tracking map of a wildfire over six days

Measuring and mapping change

Measure and map change to show the amount, percentage, or rate of change in a place. You calculate the difference in the amount of a category or in the value of a numeric attribute, and display the features based on these values.

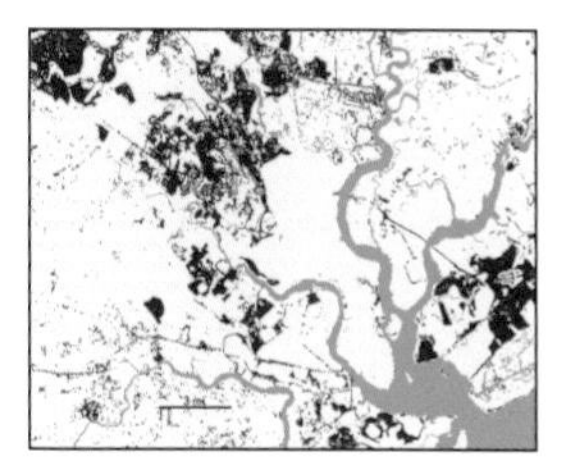

Map showing change in forest cover after a hurricane

COMPARING METHODS

Method	Type of Change	Time pattern	Pros	Cons
Time series	Movement or change in character	Trend Cycle Before and after	Strong visual impact if change is substantial; shows conditions at each date/time	Readers have to visually compare maps to see where, and how much, change occurred
Tracking map	Movement	Trend Cycle Before and after	Easier to see movement and rate of change than with time series, especially if change is subtle	Can be difficult to read if more than a few features
Measuring change	Change in character	Trend Before and after	Shows actual difference in amounts or values	Doesn't show actual conditions at each time; change is calculated between two times only

CHOOSING A METHOD

Use time series if you want to show snapshots for two or more times—either movement or change in character.

Use a tracking map if you want to show feature movement between two or more times, or over a recurring period.

Measure change if you want to show the calculated difference in an attribute of a place between two time periods.

Creating a set of time series maps is similar to making maps to show where features are or where the most and the least is, discussed in chapters 2 and 3. Because you're making a map for each of several dates or times, however, you need to consider how many maps to create, and the range of values on the maps.

You can use a time series to show change in location, or change in the magnitude or character of features.

SHOWING CHANGE IN LOCATION

A time series is effective for showing the patterns of movement if you're tracking many individual features—such as calls to 911—over time. While you can also show this using a tracking map, having many features on a single map may make the patterns difficult to see.

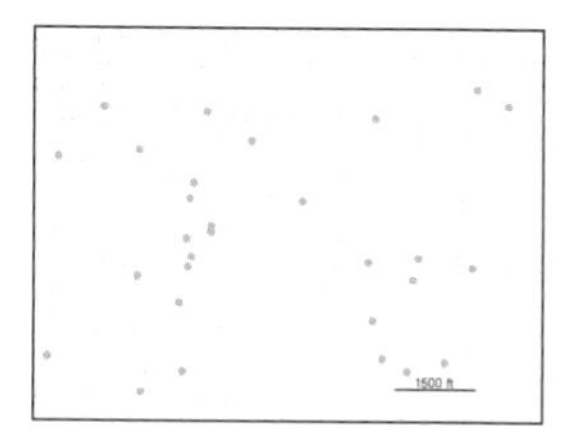

6 A.M. to noon

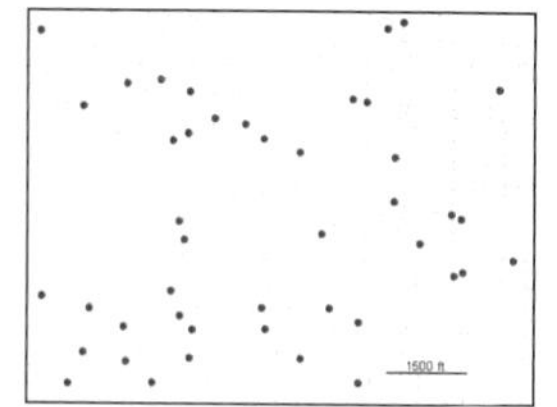

Noon to 6 P.M.

6 P.M. to midnight

Midnight to 6 A.M.

While calls to 911 during the day and evening are scattered throughout the area, late-night calls are concentrated in one area.

You can also show movement using time series if you have a few large, distinct features, such as the boundary of a wildfire or oil spill that has spread over several days.

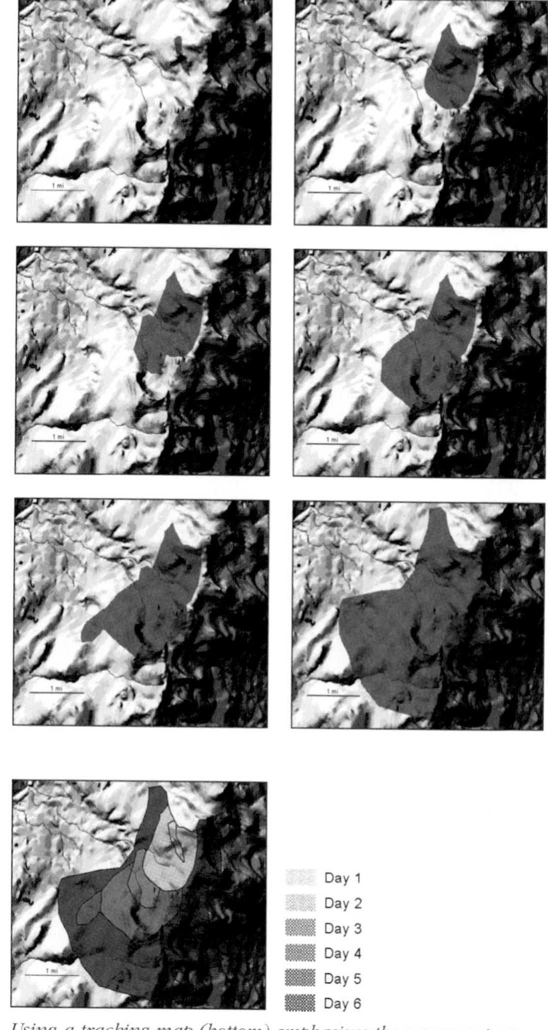

Using a tracking map (bottom) emphasizes the movement or flow of the fire, while a time series emphasizes the increase in size of the burned area.

When making your map, you'll want to include some features that are stationary, for reference. For example, when mapping crimes, you'd also draw streets on all your maps.

SHOWING CHANGE IN MAGNITUDE OR CHARACTER

A time series is particularly good for showing change in magnitude or character for discrete areas and surfaces, especially if the change is large. If the change is slight, measuring and mapping change (covered in the next section) might be more appropriate.

Change in magnitude

If you're showing change in a magnitude or quantity, you'll need to classify the values for each map. The same issues apply as discussed in chapter 3, 'Mapping the most and least'—you can create a custom classification, or use a standard classification scheme, such as natural breaks, quantile, or equal interval. However, you now have to consider the full range of values on all the maps.

One approach is to use a set of class ranges unique to each map, to best reveal the patterns on each. However, this will make it more difficult to quickly see the patterns of change between the maps—the reader will have to study each map legend to compare the change in values for each feature on each map.

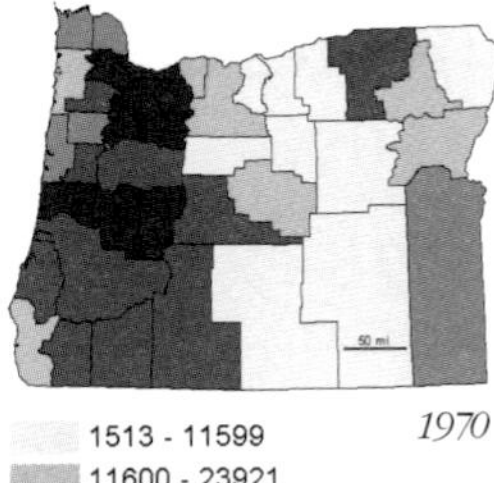

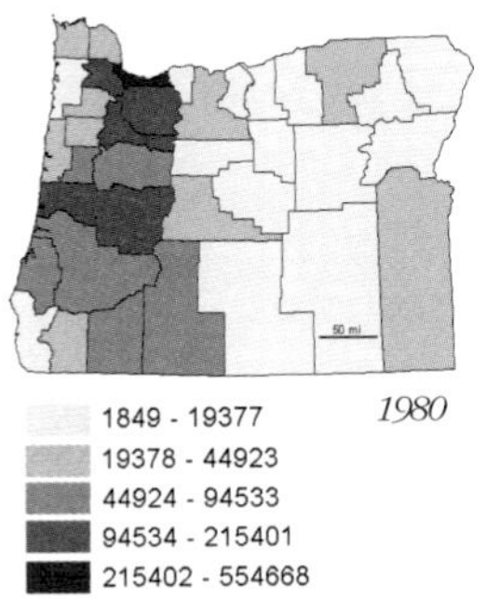

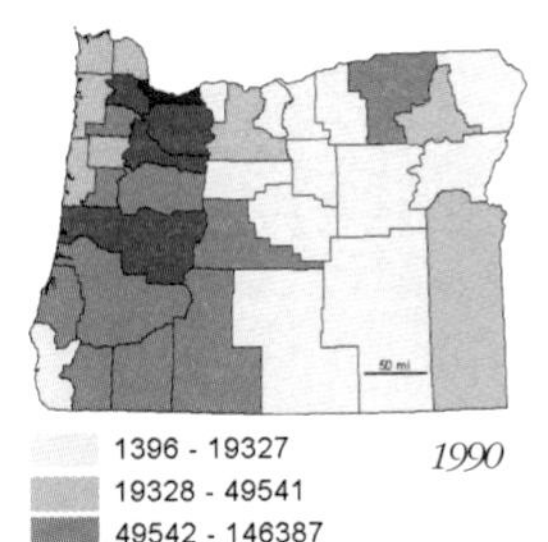

1970, 1980, and 1990 population. Using a natural breaks classification unique to each map, some counties move into a lower class between maps even as population increases, making it difficult to see quickly where growth occurred.

A more effective approach is to use a scheme applied to the full range of data values on all the maps. You can use a histogram to determine the class breaks, and manually set the class ranges to be the same for each map.

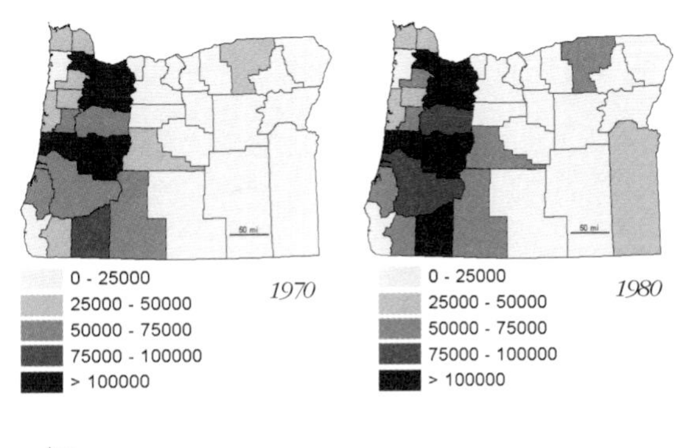

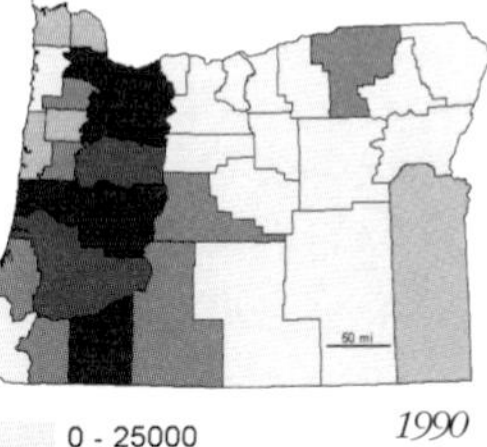

Using an equal interval classification to cover the range of values for all maps makes it clear where population growth has occurred.

Using natural breaks over the duration can reveal groupings of values in the data. If, however, the values have changed substantially over the duration, most of the values may be in one or two classes for any given map. For example, population may cluster at the low end in the earlier time periods and the high end at the later time periods.

Quantile and equal interval classification schemes are particularly useful for comparing values over time. Quantile can show whether the rank order of features remains the same or changes over time (even though the actual magnitude of values may increase or decrease for any given class). Equal interval can show whether the number of features in any given class is increasing or decreasing over time.

Change in character

When mapping change in the character of features, you may find the way the categories are defined differs between dates. This may especially be true if you have historical data, or data collected from various sources and originally used for different purposes. You can either use the existing categories for each map, or use categories that apply across all the maps.

Using the existing categories is most accurate, although it may make it harder to compare the maps. You might use this approach if the categories for the various dates are similar or if only a few categories have changed. Using different shades of the same color for categories that are similar helps make the patterns easier to see. You'll want to include notes on the map to help readers understand the similarities and differences between the categories.

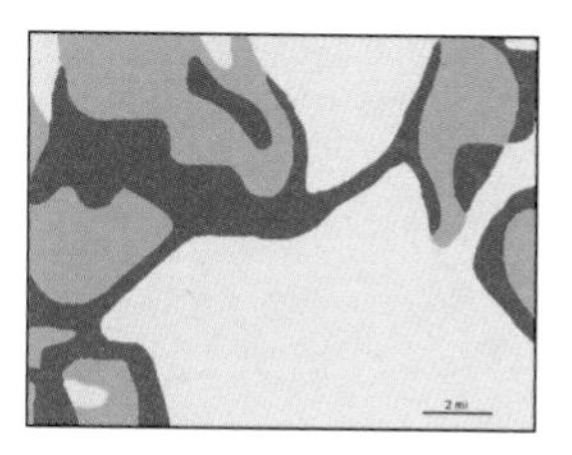

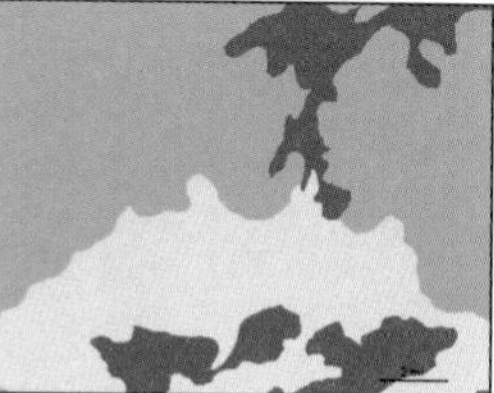

Brush
Non-timber areas
Merchantable timber

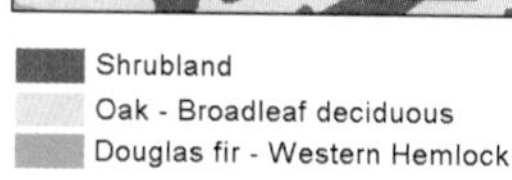

Land cover in 1914 and 1988, using the original categories

If you're mapping a few categories, you may be able to generalize them into one set of category definitions. In some cases, the categories for one date may be more detailed and can easily be generalized to the categories for the other date. For example, one map may show various categories of forest—Douglas-fir, western hemlock, ponderosa pine, and so on—which can be grouped into the more general "Conifer forest" category on the other map. In other cases, the definitions of the categories may vary only slightly, and you can broaden the definitions to include the categories on all the maps. In either case, you'll want to include text on the map making it clear how categories have been generalized, and any differences between the definitions.

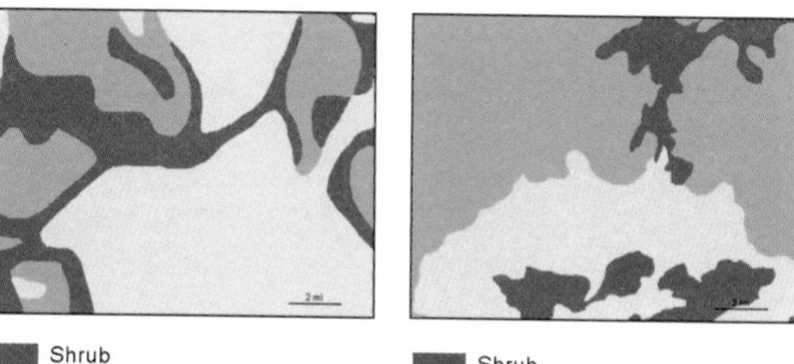

Land cover in 1914 and 1988, using generalized categories

THE NUMBER OF MAPS TO SHOW

If you have a range of dates to choose from, you can decide how many maps to show. Showing fewer maps, farther apart in time, may make the change in values easier to see. Showing more maps closer together in time may reveal patterns that are missed when using fewer maps.

It's difficult to compare more than five or six maps. Displaying fewer maps may reveal the patterns without overwhelming the reader with information.

Calls to 911

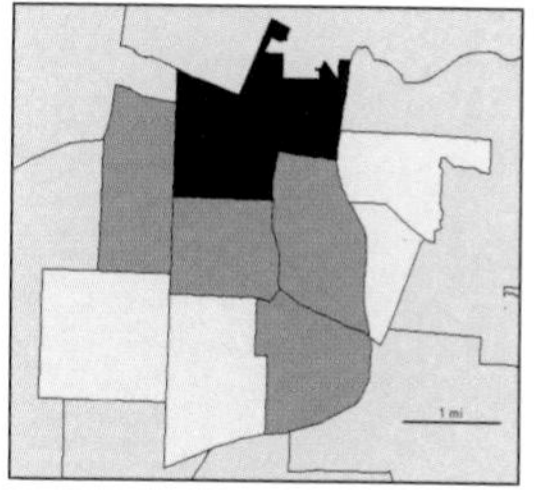

January

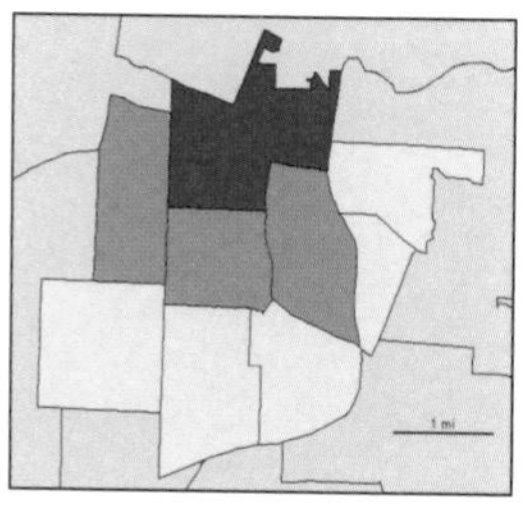

February

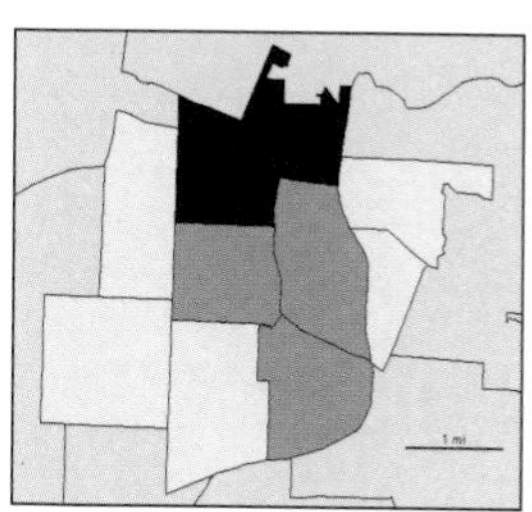

March

April

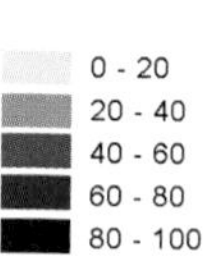

LOOKING AT THE RESULTS

Displaying tables and charts along with your maps can help show change. For example, you could create a table for each map showing the amount or percentage of each category. A bar chart (showing amounts) or pie chart (showing percentages) can also help show the changing character of a place.

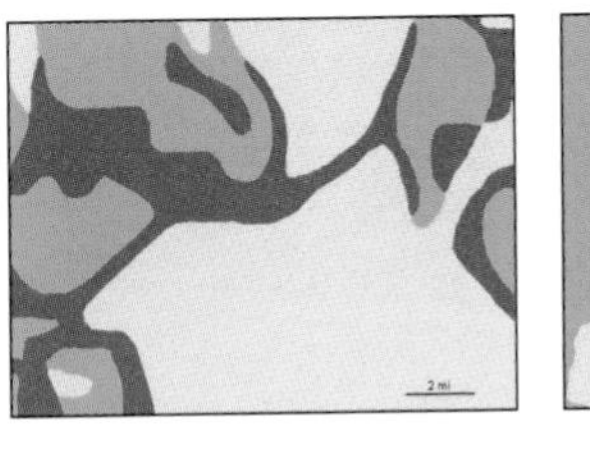

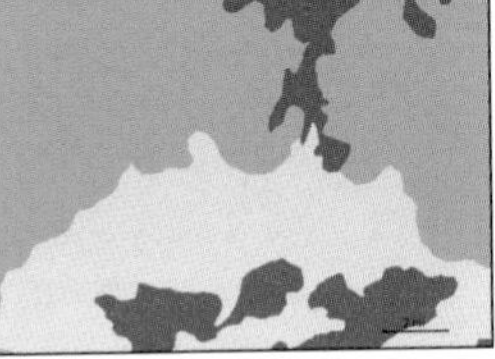

Shrub
Broadleaf forest
Conifer forest

Category	Acres	Percent
Brush	102	22
Woodland	248	53
Forest	119	25

Category	Acres	Percent
Brush	80	17
Woodland	146	31
Forest	243	52

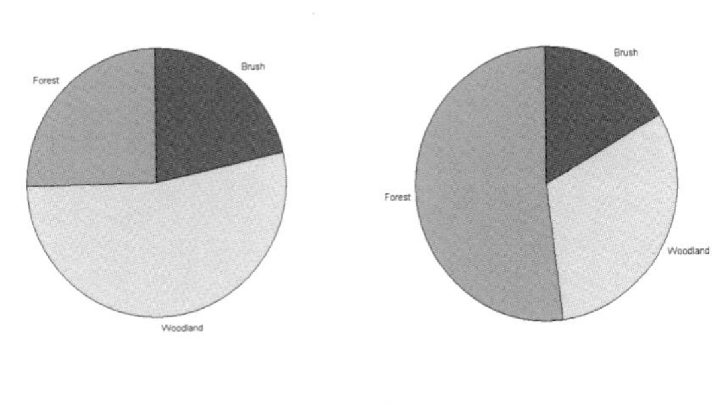

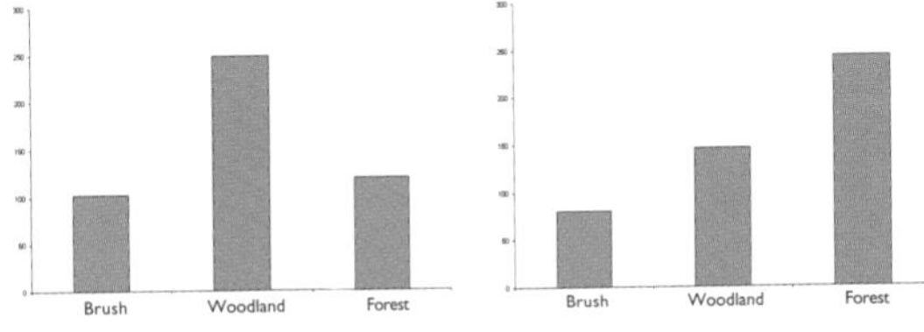

Land cover in 1914 (left column) and in 1988. The amount of woodland decreased and forest increased. While the total amount of shrub remained about the same, the maps show that it increased in some places and decreased in others.

You can also combine the tables by using the category field to join them, and create charts comparing the amount of each category at each date.

	Acres (1914)	Percent (1914)	Acres (1988)	Percent (1988)
Brush	102	22	80	17
Woodland	248	53	146	31
Forest	119	25	243	52

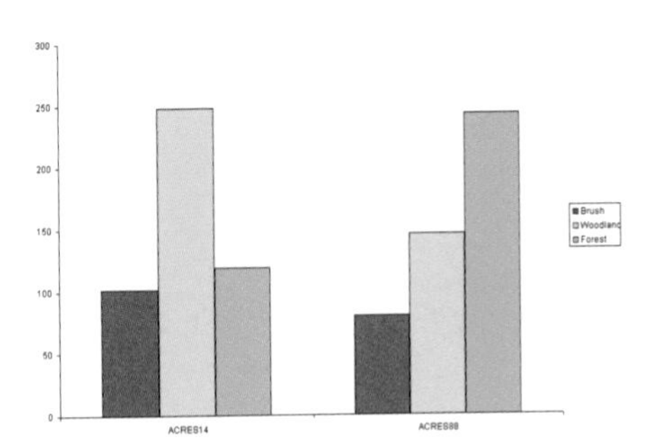

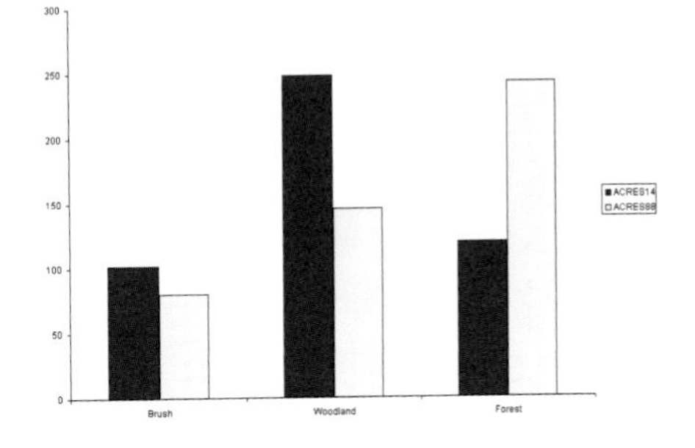

The top chart emphasizes the composition of area at each time; the bottom chart emphasizes the change in each category type.

A tracking map shows the position of a feature or features at several dates or times. It's particularly useful for showing incremental movement of discrete features (such as a hurricane or the borders of a fire) or events representing geographic phenomena (such as crime hot spots).

MAPPING INDIVIDUAL FEATURES

To map the movement of individual features represented as points, such as a hurricane or a truck, you draw each feature at each date or time. If you have several features, you can color code them to distinguish them.

If you want to emphasize the path the feature followed, draw a line to connect each date or time. The shorter the interval, the closer the line will represent the actual path the feature took. Using fewer dates or times with a larger interval generalizes the path, and may not show quick changes in direction.

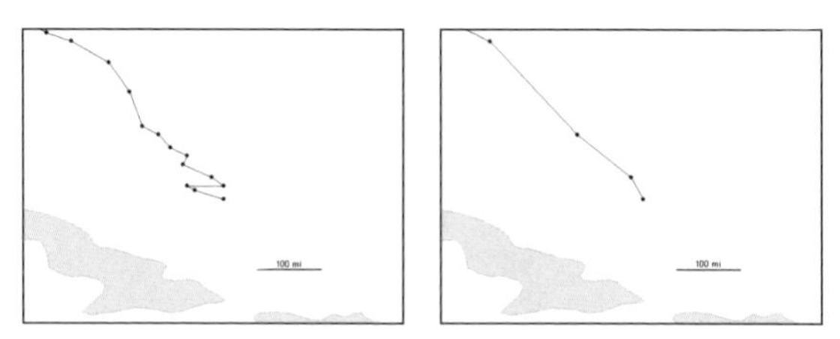

Hurricane locations at three-hour intervals (left) and six-hour intervals; with a longer interval, some of the rapid changes in direction disappear.

You show the change in character or magnitude of the features using different colors or symbols. To show change in magnitude—the wind speed of a hurricane, for example—use graduated colors or symbols, or proportional symbols.

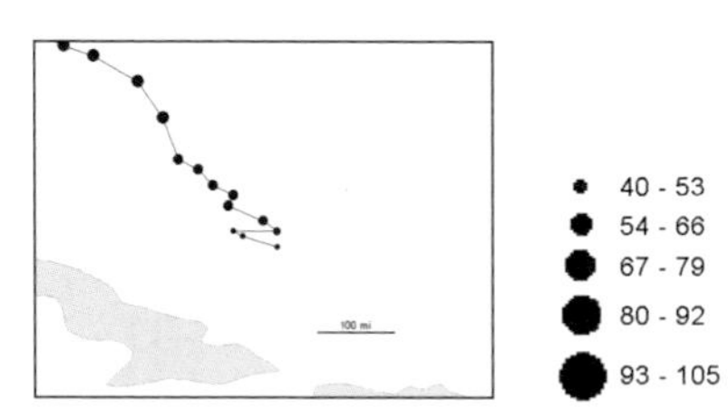

Hurricane locations mapped by maximum wind gusts

To show the rate of change, map the points at equal time intervals. For example, if you mapped the location of a hurricane every hour, you'd see where it was speeding up or slowing down. To calculate the rate of change, draw a line between the locations at each date or time, measure the length of the line, and divide by the elapsed time. You could then display the rate by drawing the lines using graduated colors or symbols. This is a generalized rate, since the feature may not have moved in a straight line.

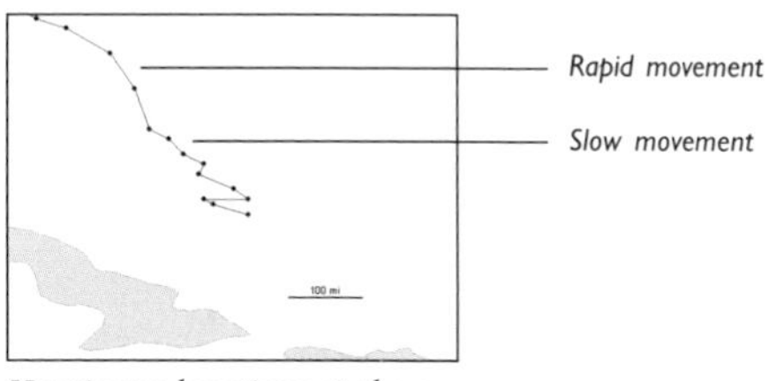

Hurricane locations at three-hour intervals—the farther apart the locations, the more rapid the movement.

MAPPING LINEAR FEATURES

To map the movement of linear features, draw them using different colors for each date or time, or label each.

Linear features are often mapped before and after an event. Examples of this are a shoreline before and after a storm, or a stream channel before and after a flood.

You can use colors or symbols to show change in the character or magnitude of the feature. For example, you might use various colors to show the change in shoreline type before and after a hurricane, or graduated line symbols to show the change in traffic volume for a road before and after realignment.

MAPPING CONTIGUOUS FEATURES

To map the movement of a contiguous feature represented as an area—such as a wildfire or an oil spill—draw the boundaries of the area at each time or date. Alternatively, shade the areas using different colors or patterns to distinguish them. You'd draw just the boundaries, without any shading, if you want to show other features under the area, such as the pre-existing land cover under the area covered by a fire.

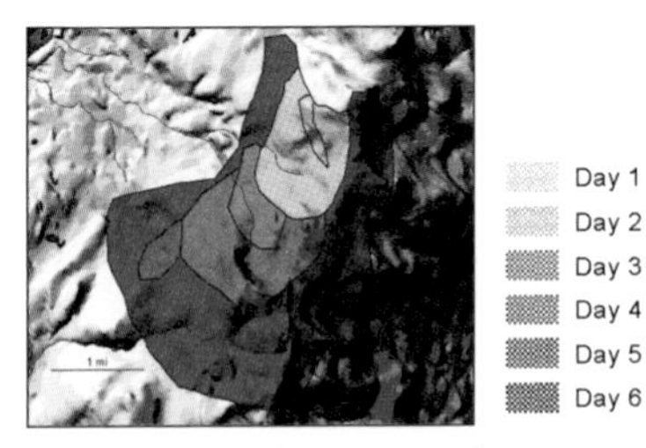

Spread of a wildfire over six days

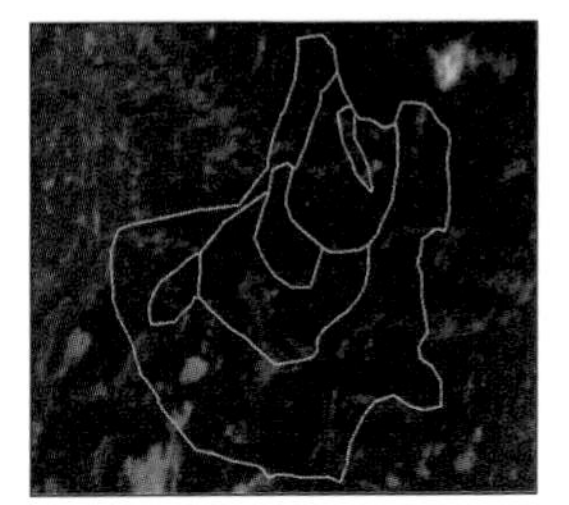

Mapping just the fire boundaries lets you see the pre-existing land cover under the burned area.

You can also shade the area based on change in character. For example, you could use graduated colors to show the average fire temperature at each date or time.

To calculate the change in areal extent over elapsed time—for example, for a fire, acres burned per hour—divide the areal extent by the period it covers. Dividing the area burned in one day by 24 hours would give you the average acres burned per hour.

By mapping an area at equal time intervals, you can see the rate of change. For example, if you map the boundaries of a fire every six hours, you can see where and when the spread of the fire is speeding up or slowing down.

MAPPING EVENTS

To show a trend in the movement of a phenomenon represented by discrete events, draw the events using a different color for each time period. For example, you might map assaults color-coded by the month in which they occurred to see if there are more assaults in an area one month, and fewer another month. That might indicate a change in the location of gang activity.

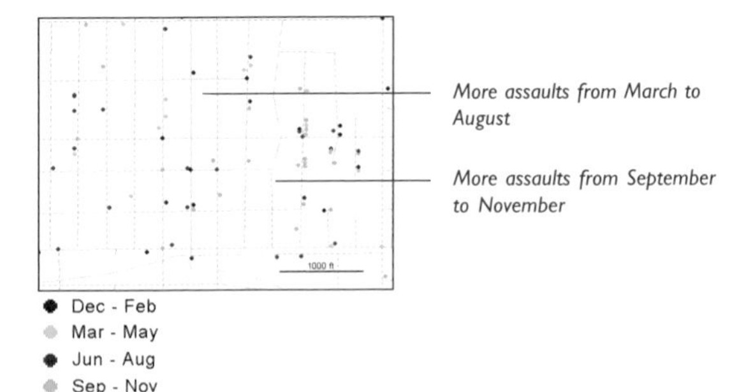

To show movement before and after an action or event, draw the events occurring before the target date in one color, and the events after in another color. For example, you'd draw narcotics arrests before a police crackdown in a neighborhood in blue, and arrests after in red, to see where drug-dealing activity moved.

To show movement over a cycle, color code the events based on the period they fall within. For example, you might draw calls to 911 color-coded by whether they occurred during the morning, afternoon, evening, or night. Showing more than six periods on a map will make it difficult for readers to pick out the patterns.

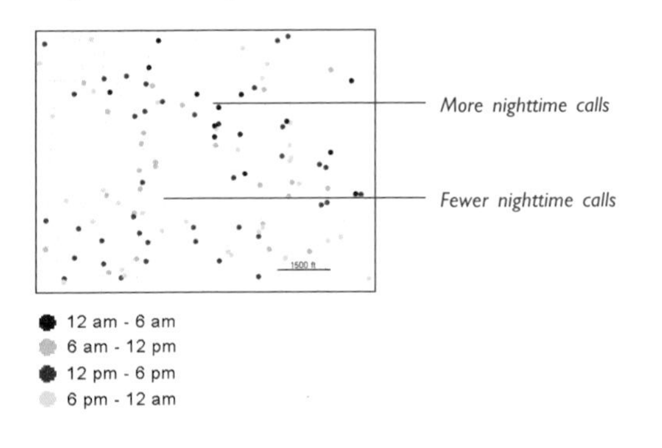

If there are multiple occurrences at each location (such as several calls to 911 from a single address), you can use pie charts to show the percentage of events occurring during each period. For example, you could show the percentage of calls occurring at each location, over the period of record, in the morning, afternoon, evening, and night. That makes it easier to see how many calls were made during each period—otherwise, a separate symbol is drawn for each call from an address, so the symbols are drawn on top of each other, obscuring the patterns.

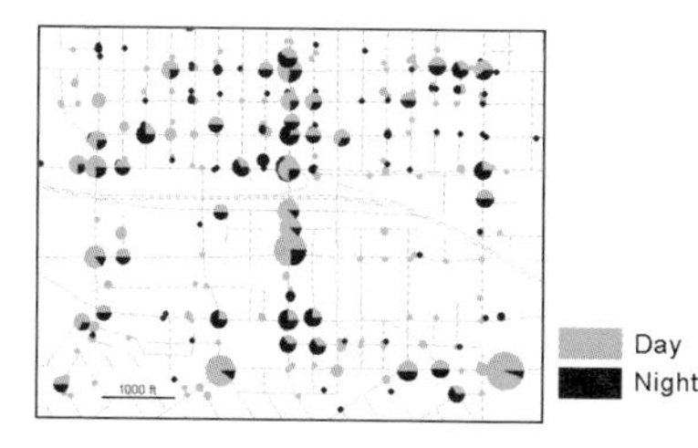

With any of these methods, you can also show the events based on magnitude. For example, you can use a graduated or proportional symbol to show the number of calls to 911 from each address. This may give you a better sense of the actual patterns over time.

LOOKING AT THE RESULTS

Showing reference features on your tracking map can start to suggest possible causes for the behavior of the features or the phenomena. For example, when mapping the path of a hurricane, you can see that the wind speed is greater over water, and much less over land, because the hurricane gets its strength by picking up moisture over the ocean.

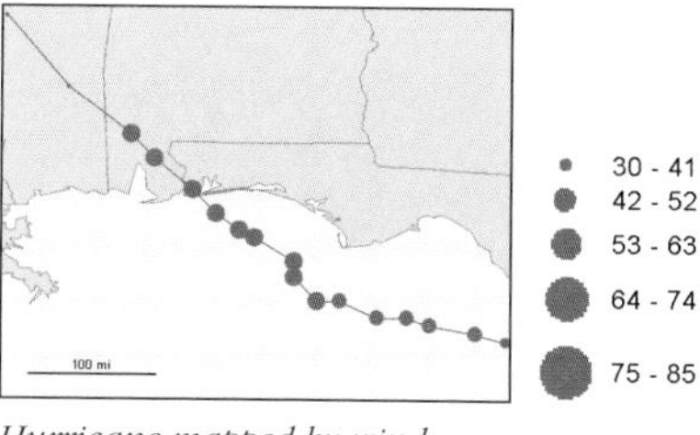

Hurricane mapped by wind speed in miles per hour

On the map below showing fire boundaries drawn on top of terrain, you can see that for two days the fire stalled at the ridge, but the next day spread rapidly, crossing over the ridge. Factors such as a change in weather conditions may have led to the expansion.

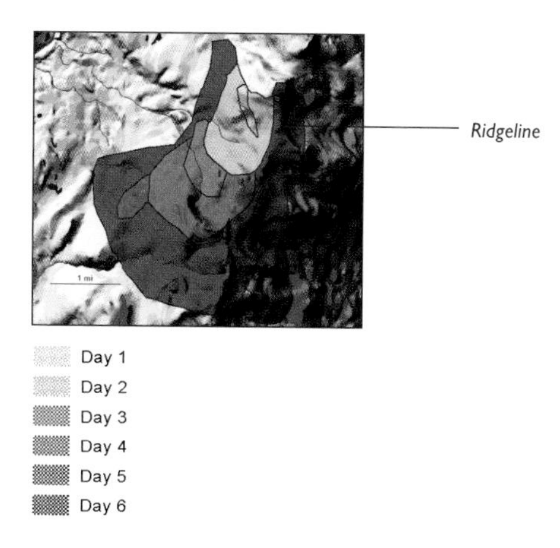

To measure and map change, you calculate the difference in value between two dates, and map features based on this value. The value can be either an amount, a percentage, or the rate of change.

You can calculate change for discrete features, data summarized by area, continuous categories, or continuous numeric values.

DISCRETE FEATURES

With discrete features, the values you're mapping are stored in the layer's data table—one value for the beginning date or time and one for the ending date or time. You subtract the beginning value from the ending value and assign the difference to a new column in the table. For example, you'd calculate the change in sales for each store by subtracting 1997 sales from 1998 sales.

Calculating change as a percentage lets you compare the relative, rather than absolute, change in values. For example, you could calculate the percentage change in sales for stores between two years to see where sales are growing fastest. To do this, divide the difference in value (calculated as shown above) by the beginning value and multiply by 100.

Making a map

For individual locations, use graduated colors or symbols to represent the calculated value. For example, you might map stores by the percentage change in sales, year over year, using graduated point symbols. Alternatively, you could display a chart at each location showing a trend line.

Change in magnitude for linear features often involves measuring flow along the features. For example, you might map roads by the change in year-over-year traffic volume, or streams by change in pollution levels for each month. Use graduated line symbols to map the change values. If the flow is measured at sample locations, such as water quality in a stream, you can use a point symbol to represent the change.

For areas, such as parcels, use graduated color to show the change in value for each. For example, mapping the percentage increase in assessed value of each parcel using graduated colors would show you which neighborhoods had the greatest change in property value during the period.

DATA SUMMARIZED BY AREA

Calculating change for data summarized by area is the same as calculating it for discrete features—the values you're mapping are stored in the layer's data table; you subtract the beginning value from the ending value, and assign the difference to a new column in the table. For example, you'd calculate the change in population for each county between 1970 and 1990 by subtracting the 1970 population from the 1990 population.

County FIPS	1970 Pop	1990 Pop	Difference
41049	4465	7625	3160
41061	19377	23598	4221
41057	18034	21570	3536
41021	2342	1717	-625
41067	157920	311554	153634

To see which counties had the greatest relative growth, you'd calculate the percentage change by dividing the difference in population by the population in 1970, and multiplying by 100.

County FIPS	1970 Pop	1990 Pop	Difference	% Change
41049	4465	7625	3160	71
41061	19377	23598	4221	22
41057	18034	21570	3536	20
41021	2342	1717	-625	-27
41067	157920	311554	153634	97

Making a map

Once you've calculated the change value, you can map the features based on the value. This is the same as mapping amounts or ratio values, as discussed in chapter 3, 'Mapping the most and least.' Use graduated colors to shade each area based on the change value.

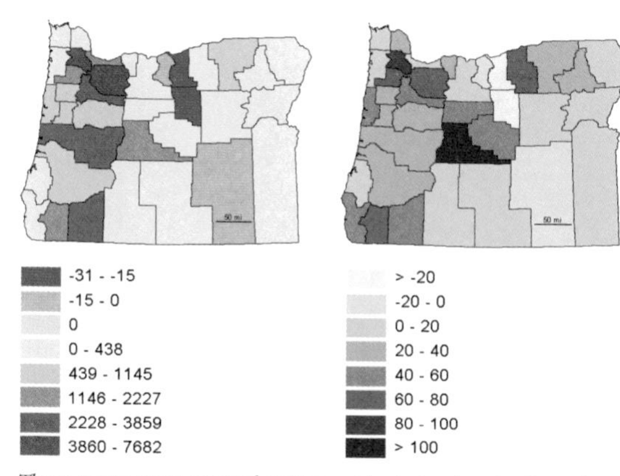

These maps were created using graduated colors to show the rate of population change (left map) and percentage change.

To show the change in value for each area for two or more periods, create bar charts showing the value for each period. You can place the chart inside each area, or create a separate bar chart that shows the values for each area side by side. Using charts works best for a few areas and a few time periods. Otherwise, it's hard for readers to see the trends.

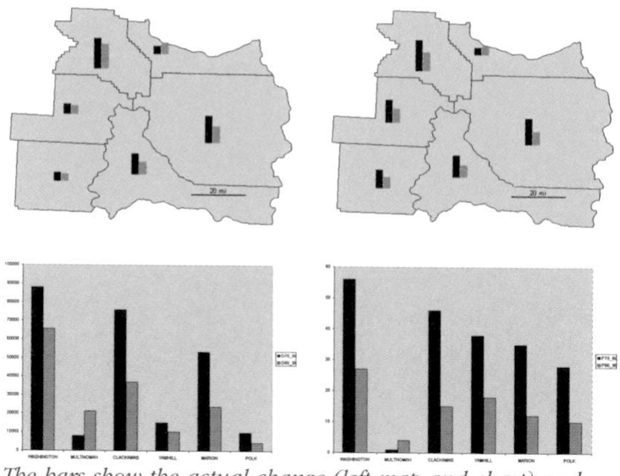

The bars show the actual change (left map and chart) and percentage change in population from 1970 to 1980 (dark blue) and 1980 to 1990 (light blue).

To show change for several dates or times, you can create a trend line chart. You can create a chart for each area, and display it in each, or create a single chart for all areas and display it in the legend. For example, you could show the population for each county every five years between 1970 and 1990.

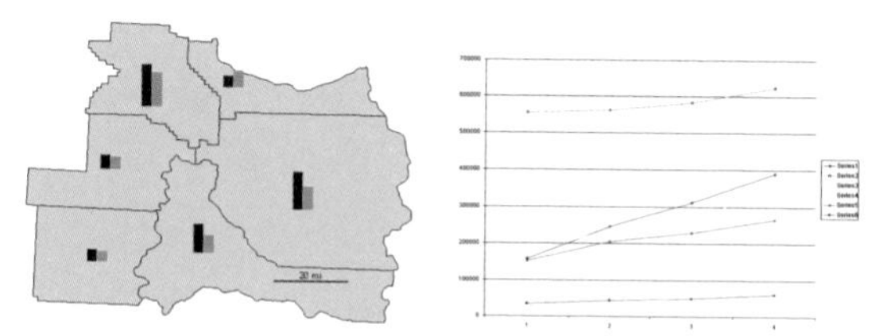

A trendline shows the relative population, as well as the growth, of each county.

Mapping negative values

After you've calculated the change, some features may have negative values. For example, population may have actually decreased in some counties. When creating a map, you can set the class ranges and symbols so that negative values are drawn using one color, and positive values in a contrasting color. Readers can then quickly see where values have increased or decreased.

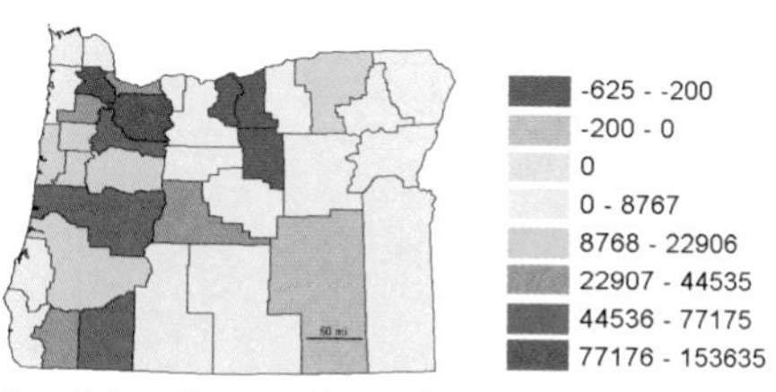

Population decreased over the period in counties shaded blue.

What if the boundaries have changed?

If you have historical data, the boundaries of the areas you're mapping may have changed. Small changes may not change the patterns on the map, especially if you're mapping a large region compared to the size of the individual areas. If the changes are minor, you can draw the original boundaries on the map using a dashed line, or a lighter shade, and include these symbols in the map legend. If the changes are substantial, you may need to find another way to map the attributes you're interested in. For example, if you're mapping the total crimes within several neighborhoods for which the boundaries have changed, you could create a continuous surface based on the center point of each neighborhood, for both time periods, and calculate the change between the surfaces (see chapter 4, 'Mapping density').

CONTINUOUS CATEGORIES OR CLASSES

Calculating change for continuous categories, such as land cover, involves combining two layers, one for each date or time.

To calculate change in categories or continuous classes, you can create a map of only the areas that have changed; calculate the change in areal extent for each category, and show it in a table or chart; or calculate the amount of change from each category to each of the other categories, and present the results in a map, table, or chart.

Creating a map of areas that have changed

You can do this using either vector or raster data, as with finding which categories are inside an area (discussed in chapter 4, 'Mapping density').

Using vector data, you overlay the areas to create a new layer containing the category codes for both dates. You then select those features for which the codes are not equal—that is, the category at the first date is different than that at the second date.

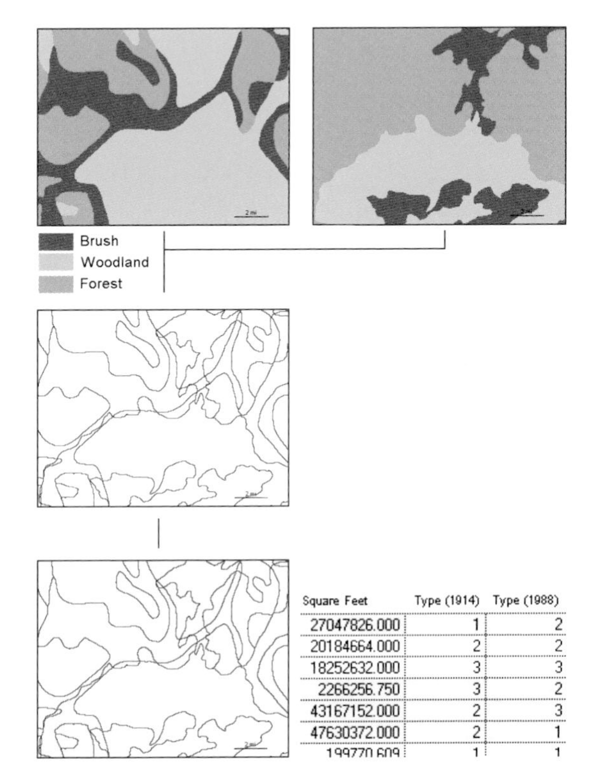

Square Feet	Type (1914)	Type (1988)
27047826.000	1	2
20184664.000	2	2
18252632.000	3	3
2266256.750	3	2
43167152.000	2	3
47630372.000	2	1
199770.609	1	1

Using raster data, you create a new layer by comparing the cell values of the two input layers. The GIS checks each pair of cells, and tags the ones for which the codes are not equal. These are the cells that have changed.

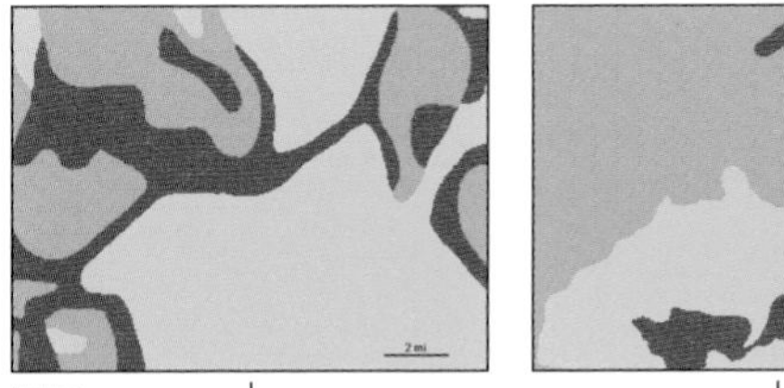

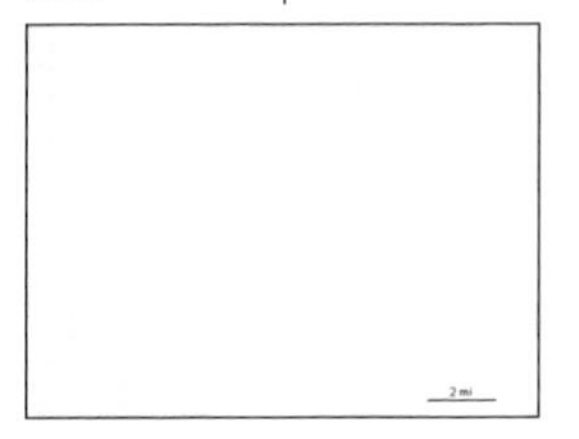

With either type of data, once you've created the layer showing the areas that have changed, you can map it along with the original features, shaded by category.

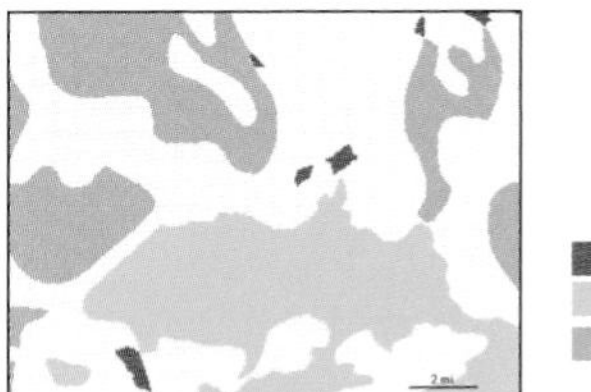

Calculating change in areal extent for each category

To calculate the change in extent by category, sum the area of each category for each date. Then, join the tables using the category code. Calculate the change by subtracting the area for the original date from the area at the second date and assigning it to a new column. You can also calculate the percentage change by dividing this value by the area for the original date and multiplying by 100.

Category	Acres
Brush	102
Woodland	248
Forest	119

Category	Acres
Brush	80
Woodland	146
Forest	243

Category	Acres (1914)	Acres (1988)	Difference	% Change
Brush	102	80	-22	-22
Woodland	248	146	-102	-41
Forest	119	243	124	104

This method doesn't create a layer showing the changed area. You'll need to display the information by showing a table or chart of the results, along with the maps from both dates.

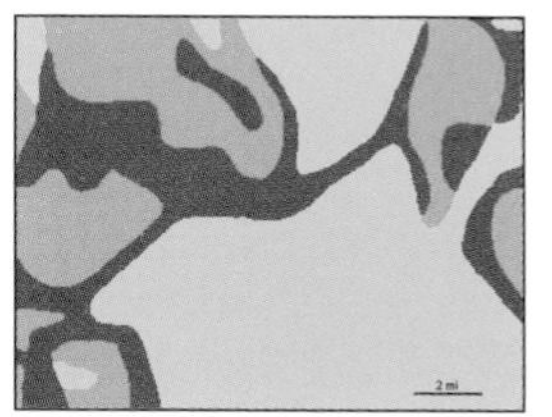

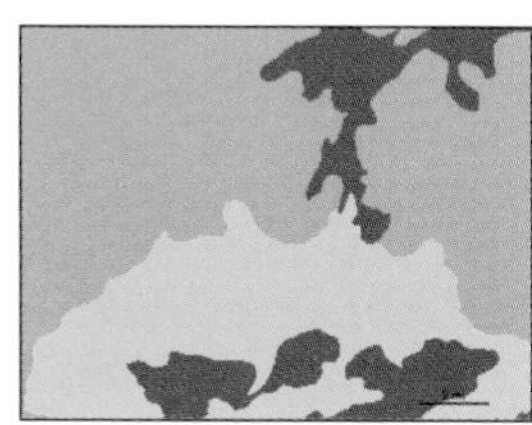

Brush
Woodland
Forest

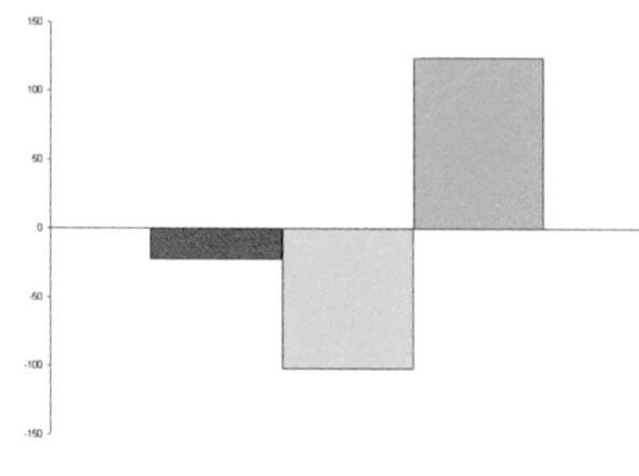

Change in acres for each land cover type

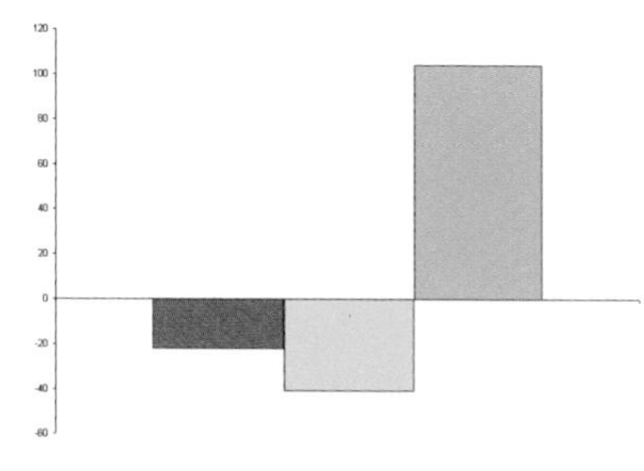

Percentage change in each land cover type

Calculating change from each category to every other category

As with creating a map of the area that's changed, you can use either vector or raster data to calculate how much of each original category has changed to each new category.

Using vector data, you overlay the two layers as described in chapter 5, 'Finding what's inside.' You then create a frequency table using the category codes for both dates and summing the area attribute. The result is a table that shows the areal extent of each combination of "from" category and "to" category. You then assign a unique value to each combination, join the frequency table to the layer's data table, and map each area based on its new code.

From-To Code	Acres
11	1135
12	4213
13	21083
21	17799
22	31117
23	15158
31	1644
32	2436
33	26611

With raster data, the GIS compares the two input feature sets and creates a map and matrix showing the area of each code combination. The map assigns a value to each cell based on its from–to combination.

	1	2	3
1	1135	4213	21083
2	17799	31117	15158
3	1644	2436	26611

Showing each combination of codes works best with six codes at most. With more, the map becomes difficult to read.

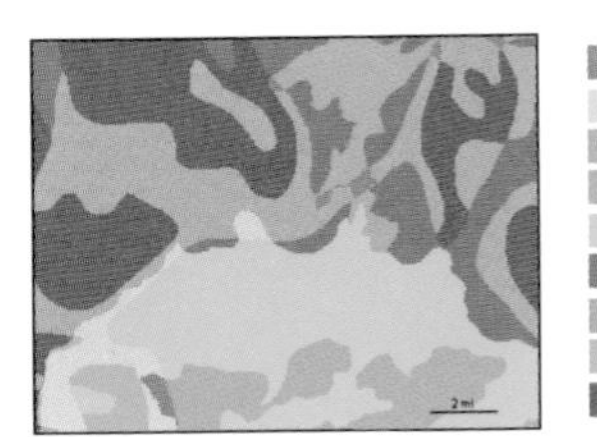

- Brush (no change)
- Brush to Woodland
- Brush to Forest
- Woodland to Brush
- Woodland (no change)
- Woodland to Forest
- Forest to Brush
- Forest to Woodland
- Forest (no change)

Even with just three categories of land cover, the resulting map is complex.

The way in which you symbolize the code combinations depends on whether you're more interested in what it changed from, or what it changed to. Using shades of the same color for the "from" categories emphasizes what category each area was originally.

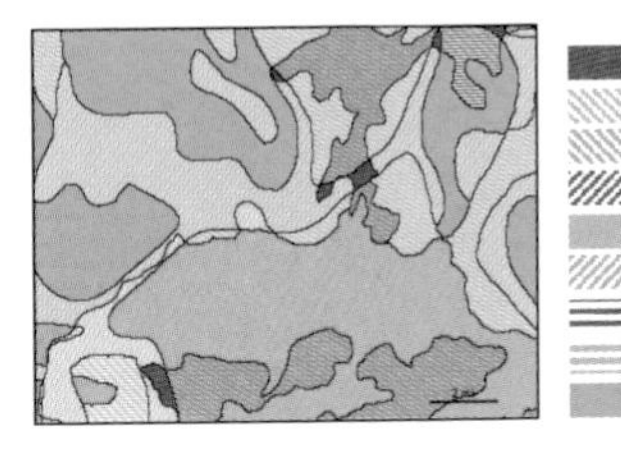

- Brush (no change)
- Brush to Woodland)
- Brush to Forest
- Woodland to Brush
- Woodland (no change)
- Woodland to Forest)
- Forest to Brush
- Forest to Woodland
- Forest (no change)

You can also use charts to display the tabular information.

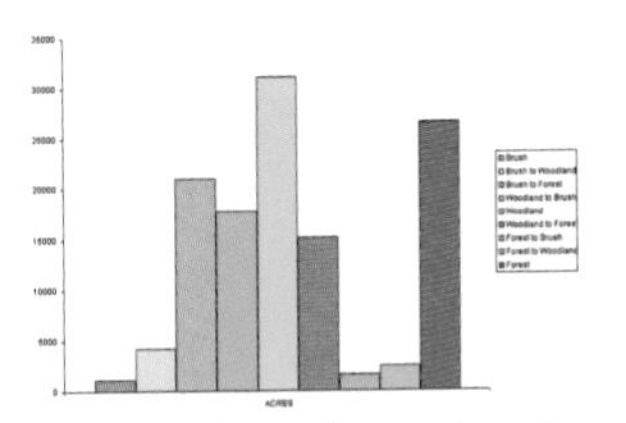

The chart shows the number of acres for each "from–to" combination.

What if the category definitions have changed?

As with using time series for categories, you may find category definitions have changed between the dates, especially if you're dealing with historical data, or with data layers that were originally intended for disparate uses. You can use the existing categories, or generalize to create categories that are the same for both dates.

Using the existing categories is more accurate. However, you may end up with a large number of combinations, making the patterns on the map more difficult to see. Generalizing can distort the data, but the patterns of change may be easier to see and understand, especially if you can generalize to create fewer categories.

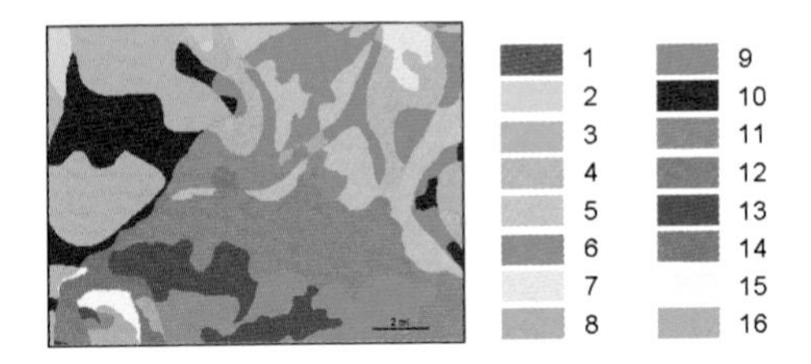

From-To Code	"From" Category	"To" Category
1	Non-timber	Douglas fir - Western Hemlock
2	Timber	Douglas fir - Western Hemlock
3	Cutover area	Douglas fir - Western Hemlock
4	Cutover area	Broadleaf deciduous
5	Timber	Broadleaf deciduous
6	Timber	Oak-Pine pasture
7	Cutover area	Oak-Pine pasture
8	Burned area	Oak-Pine pasture
9	Burned area	Broadleaf deciduous
10	Burned area	Douglas fir - Western Hemlock
11	Timber	Douglas fir - Oak
12	Burned area	Douglas fir - Oak
13	Timber	Oak woodland
14	Burned area	Oak woodland
15	Cutover area	Oak woodland
16	Cutover area	Douglas fir - Oak

In this example, the earlier forest map used land cover categories ("from" category), while the later map used vegetation types ("to" category). Mapping each combination results in 16 "from–to" codes, making for a very complex map.

CONTINUOUS NUMERIC VALUES

You may have continuous numeric values for two dates for which you want to calculate change. These might be based on density, such as a density surface of crimes per square mile or a surface of land value per square foot. Or, they may be interpolated values, such as the concentration of an air pollutant over an area, created from values at point-sampling stations.

To create a change map between two surfaces, you subtract the layers. The GIS calculates the difference between each cell on the first layer and the corresponding cell on the second layer. The resulting map shows where the most and least change occurred.

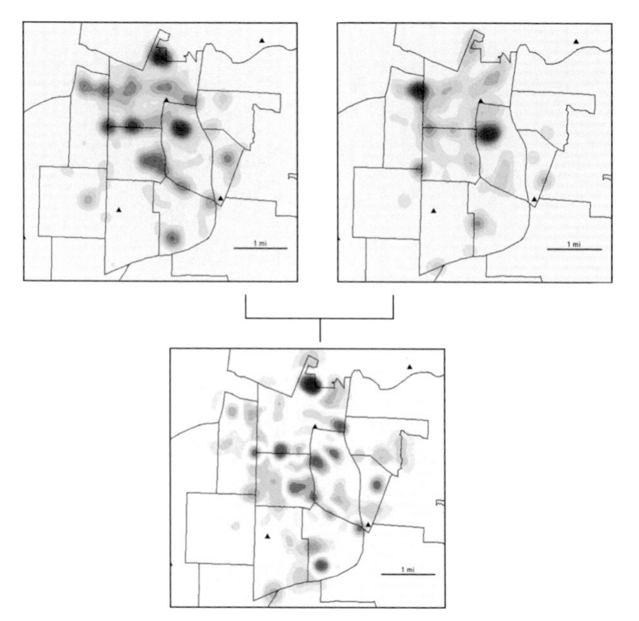

Density surfaces of calls to 911 for January (upper left) and February (upper right). By subtracting the surfaces, you can see where calls increased most (darkest orange) and decreased most (darkest blue).

Where to get more information

These books and articles will help you explore GIS analysis on your own. While this is list is by no means comprehensive, the sources here provide a good introduction to each topic.

Geographic databases and map projections

Kennedy, Melita. *Understanding Map Projections*. ESRI, 1999.

This handbook presents an overview of map projection concepts and the steps involved in projecting geographic data. It also details the map projections supported by ESRI's ArcInfo GIS.

Snyder, John P. *Map Projections—A Working Manual*. U.S. Geological Survey, 1989.

This classic work on map projections covers all the most commonly used projections, describing when they're best used, and their shortcomings. This technical manual includes the mathematical formulas used to calculate each projection.

Zeiler, Michael. *Modeling Our World—The ESRI Guide to Geodatabase Design*. ESRI, 1999.

Includes an extensive review of the various GIS data and file formats, including vector and raster data, geometric networks, and surfaces. The book describes how geographic features and attributes are stored and displayed, and how geographic databases are structured.

Making maps, using symbols, and classifying data

MacEachren, Alan M. *Some Truth with Maps: A Primer on Symbolization and Design*. Association of American Geographers, 1994.

This primer offers a succinct, practical introduction to making and using maps. It includes sections on using symbols and colors, and on data classification. The focus is on how readers perceive the information on the map.

Monmonier, Mark. *How to Lie with Maps*. University of Chicago, 1991.

A highly accessible and informative look at how maps are used and misused. Especially useful are its discussions of how data classification and use of symbols can clarify or obscure patterns in your geographic data.

Using ArcView GIS. ESRI, 1996.

This guide to using ESRI's ArcView GIS includes information on data classification, and how to create maps using different types of symbols. It also includes discussions of working with tables, and geographic query and selection.

Using ArcMap. ESRI, 1999.

A guide to using ArcMap, part of ESRI's ArcInfo GIS, for creating maps and working with tabular data. The guide includes sections on data classification, using symbols, and selecting and querying geographic features.

Measuring and analyzing geographic data

Berry, Joseph K. *Beyond Mapping: Concepts, algorithms, and issues in GIS*. GIS World Books, 1993.

Berry's collection of articles, geared toward the practitioner, covers various analysis topics, including measuring distance, overlay, and analyzing patterns. Berry also addresses how to assess the amount of error in your analysis results.

Chrisman, Nicholas. *Exploring Geographic Information Systems*. Wiley, 1997.

Chrisman describes the functions and use of geographic information systems in the context of their broader role in society. Includes an excellent discussion of ways to measure and combine geographic data, and of working with attributes.

DeMers, Michael N. *Fundamentals of Geographic Information Systems*. Wiley, 1997.

This textbook provides a comprehensive overview of GIS, covering how geographic features and attributes are stored and manipulated. Also includes extensive discussions of various geographic analysis operations.

Working with rasters and surfaces

Tomlin, C. Dana. *Geographic Information Systems and Cartographic Modeling*. Prentice–Hall, 1990.

Tomlin's definitive work provides an in-depth discussion of how information is represented using raster data, and the issues involved in working with raster data. Also includes a technical discussion of the concepts involved in raster overlay, calculating distance over a surface, and other raster operations.

Using ArcView Spatial Analyst. ESRI, 1997.

This guide to ESRI's ArcView Spatial Analyst software provides basic information on working with and displaying raster data in ArcView GIS. It includes brief descriptions of the various available raster analysis functions, including distance mapping and raster overlay.

Using ArcView 3D Analyst. ESRI, 1998.

The software guide includes an introduction to the various ways geographic surfaces are created and stored in a GIS. Also describes how to display and analyze surfaces as 3-D perspective views using ESRI's ArcView 3D Analyst.

Working with geographic networks

Chou, Yue-Hong. *Exploring Spatial Analysis in Geographic Information Systems*. Onword Press, 1997.

Chou's book provides a broad overview of spatial analysis applications. It includes a fairly extensive chapter on network analysis, with a technical discussion of the structure and requirements of a geographic network, and a description of how a GIS solves typical network analysis problems, such as routing and finding the shortest path.

Using ArcView Network Analyst. ESRI, 1997.

This guide to ESRI's Network Analyst software includes a good discussion of building a network layer for analysis, including how to specify distance or cost, assigning turns and stops, and barriers to travel. Also includes a brief chapter on using ArcView GIS to find a service area around a center.

Time and GIS

Blok, Connie, et al. "Visualization of relationships between spatial patterns in time by cartographic animation," *Cartography and Geographic Information Science*, Vol. 26, No. 2, 1999, pp. 139–151.

This article describes research on the use of computers to animate a map time series to look for relationships between geographic phenomena. The researchers' work focused on the issues involved in animating two superimposed map series. The article also includes a good list of references on time and maps.

MacEachren, Alan M. *How Maps Work*. Guilford Press, 1995.

MacEachren's extensive book on how maps are represented and perceived includes a chapter on relationships in space and time. The chapter covers the concepts behind visual analysis of change, and ways of organizing time. MacEachren also addresses the use of computers to animate a time series to look for patterns.

Monmonier, Mark. "Strategies for the visualization of geographic time series data," *Cartographica*, Vol. 27, No. 1, 1990, pp. 30–45.

This article describes various approaches for mapping change in both the location and character of geographic features, including time series, tracking maps, and computer animations.

Muehrcke, Phillip C. and Juliana O. Muehrcke. *Map Use*. JP Publications, 1992.

This highly accessible book, covering all aspects of map creation and use, includes a chapter on time and maps. Topics covered include mapping trends and cycles, and issues related to time sensitivity of geographic data.

Map and data credits

A number of people graciously contributed their work to make the examples in the book reflect the real world use of GIS for analysis.

MAP CREDITS

The following organizations and individuals created and provided the maps that appear in the 'Map gallery' section of each chapter.

Chapter 2

"Selections from Oregon's Forest Atlas"
Oregon Department of Forestry, Salem, Oregon
By Emmor Nile
Copyright © 1996 Oregon Department of Forestry

"City of North Vancouver—Pavement Management"
City of North Vancouver, British Columbia, Canada
By Trevor Fawcett

"San Diego Region Vegetation Map"
SANDAG, San Diego, California
By Sue Carnevale and John Hofmockel

"City of Irvine Zoning Map"
Community Development Department, City of Irvine, California
By Victor Kao, Chinh Nguyen, Sherry Richardson, and John Ernst
Map courtesy of the Community Development Department, City of Irvine, California

Chapter 3

"Student Truant Locations, City of Stockton"
MIS Department, GIS Division, City of Stockton, California
By Robert MacLeod

"Community Development Block Grant, Louisville, Kentucky"
City of Louisville Department of Public Works, Louisville, Kentucky
By Gregory B. Singlust

"Utilization Characteristics of Primary Health Care Centers, Ministry of Public Health, State of Qatar"
Ministry of Public Health, Doha, State of Qatar
By Dr. Noor-Amin Noorani, Dr. Abdul Rahman Al-Kuwari, Mohamed Al-Noimi, and Mr. R. Chandrasekharan
Map courtesy of the Ministry of Public Health, Doha, State of Qatar

"Southern California Edison, Customer Profile for District 32, Field Service Calls, Jan. 1 to March 31, 1997"
Southern California Edison (SCE), Rosemead, California
By James Rodriguez, Ryan Damon, and Joshua Schechter

Chapter 4

"Narcotics Arrests—All Types—1995, City of Stockton"
MIS Department, GIS Division, City of Stockton, California
By Robert MacLeod

"A Location Analysis of Secretary of State Offices in Southeast Michigan"
Southeast Michigan Council of Governments, Detroit, Michigan
By George Janes, John Baran, and Ann VanSlembrouck
Copyright © 1996 SEMCOG

"Fish Densities in Lake Norman, September, 1994"
Duke Power Company, Huntersville, North Carolina
By Tim Leonard and Don Degan

"Selections from Oregon's Forest Atlas"
Oregon Department of Forestry, Salem, Oregon
By Emmor Nile
Copyright © 1996 Oregon Department of Forestry

Chapter 5

"Gold Fork Watershed Subbasin Analysis"
Boise Cascade Corporation, Boise, Idaho
By Brian Liberty

"The Corridor Major Investment Study"
Parsons Brinckerhoff, Orange, California
By Dana Brown, David Freytag, and Kevin Keller

"Nellis Air Force Base—Airport Environs"
Clark County GIS Management Office, Las Vegas, Nevada
By Ralph Spear
Copyright © 1996 Ralph Spear

"Formations Végétales et Domaine Forestier National de Madagascar"
Conservation International, Washington, D.C.
By Lata Iyer, S. Olivieri, Chris Rodstrom, Andrew Waxman, and Gaston Razafindramboa

Chapter 6

"Prince William County Department of Fire and Rescue Unit Travel Time Zones"
Prince William County Department of Fire and Rescue, Prince William, Virginia
By David J. Simms, GIS Administrator

"Land Use Codes, Seattle Public Library, Library Map Series"
City of Seattle Corporate GIS
Seattle Engineering Department, Seattle, Washington
By Steve Beimborn and Jarek Domanski
Copyright © 1997

"Customer Proximity to Software Retailers, Metropolitan Seattle"
Decision Support Services, Brooklyn, New York
By Joe Lewis, Duncan Bruce, and Chris Nimmo
Copyright © 1996–1997 Decision Support Services, Inc.

"City of Winston-Salem—Fire Station First Response"
City of Winston-Salem, North Carolina
By Tim Lesser

Chapter 7

"Emaralda Marsh Vegetation Change Analysis (1941–1987)"
St. Johns River Water Management District, Palatka, Florida
By M. Tepera, G. Dambek, E. Gisondi, G. Greenburg, J. Marburger, and W. VanSickle
Map courtesy of the St. Johns River Water Management District, Palatka, Florida

"Spatial and Temporal Trends in California Motor Vehicle Activity and Exhaust Emissions"
California Air Resources Board, Motor Vehicle Analysis Branch, El Monte, California
By Michael T. Benjamin, Daniel Hawelti, and Mark Carlock
Copyright © 1997 California Air Resources Board

"Landfalling Hurricanes (1886–1996) By Month"
South Florida Water Management District (SFWMD), West Palm Beach, Florida
By Ed Biggs, Terri Bennett, Eric Swartz, Geoff Shaughnessy, Chris Burns, and John Neuharth

"Modeling and Mapping Solar Insolation in the Dry Valleys of Antarctica: Hours of Direct Sunshine Over a 24-hour Period"
Colorado State University, Fort Collins, Colorado
By Keith Croteau and Dr. Andrew Fountain
Copyright © 1997 Keith Croteau

DATA CREDITS

The following organizations and individuals provided GIS datasets used to create the map examples throughout the book.

Metro Regional Services
Data Resource Center
600 NE Grand Ave.
Portland, OR 97232
503-797-1742
drc@metro.dst.or.us

Tualatin Valley Fire and Rescue
Aloha, Oregon
Skip Kirkwood, Chief, Infomatics and Quality Improvement Unit

Much of the data representing natural and cultural themes has been provided courtesy of the State Service Center for Geographic Information Systems:

State Service Center for Geographic Information Systems
155 Cottage St. NE
Salem, OR 97310
503-378-2166
data@sscgis.state.or.us

Index

Symbols

A

B

C

D

E

F

G

H

I

J

L

T

U

V

W

Z